Advance in Mechanical and Thermal Characterization of Polymer Composites

Advance in Mechanical and Thermal Characterization of Polymer Composites

Editor

SD Jacob Muthu

Basel • Beijing • Wuhan • Barcelona • Belgrade • Novi Sad • Cluj • Manchester

Editor
SD Jacob Muthu
Faculty of Engineering &
Applied Science
University of Regina
Regina
Canada

Editorial Office
MDPI
St. Alban-Anlage 66
4052 Basel, Switzerland

This is a reprint of articles from the Special Issue published online in the open access journal *Polymers* (ISSN 2073-4360) (available at: www.mdpi.com/journal/polymers/special_issues/D03UN7657W).

For citation purposes, cite each article independently as indicated on the article page online and as indicated below:

Lastname, A.A.; Lastname, B.B. Article Title. *Journal Name* **Year**, *Volume Number*, Page Range.

ISBN 978-3-0365-8653-3 (Hbk)
ISBN 978-3-0365-8652-6 (PDF)
doi.org/10.3390/books978-3-0365-8652-6

© 2023 by the authors. Articles in this book are Open Access and distributed under the Creative Commons Attribution (CC BY) license. The book as a whole is distributed by MDPI under the terms and conditions of the Creative Commons Attribution-NonCommercial-NoDerivs (CC BY-NC-ND) license.

Contents

About the Editor vii

Preface ix

Ibraheem A. Abdulganiyu, Oluwasegun. E. Adesola, Ikechukwuka N. A. Oguocha and Akindele G. Odeshi
Dynamic Impact Properties of Carbon-Fiber-Reinforced Phenolic Composites Containing Microfillers
Reprinted from: *Polymers* **2023**, *15*, 3038, doi:10.3390/polym15143038 1

Jibing Chen, Bowen Liu, Maohui Hu, Qianyu Shi, Junsheng Chen and Junsheng Yang et al.
Research on Characterization of Nylon Composites Functional Material Filled with Al_2O_3 Particle
Reprinted from: *Polymers* **2023**, *15*, 2369, doi:10.3390/polym15102369 26

Raja Venkatesan, Krishnapandi Alagumalai, Chaitany Jayprakash Raorane, Vinit Raj, Divya Shastri and Seong-Cheol Kim
Morphological, Mechanical, and Antimicrobial Properties of PBAT/Poly(methyl methacrylate-*co*-maleic anhydride)–SiO_2 Composite Films for Food Packaging Applications
Reprinted from: *Polymers* **2022**, *15*, 101, doi:10.3390/polym15010101 36

Dae-Hee Kim, Jeong-Hyeon Kim, Hee-Tae Kim, Jeong-Dae Kim, Cengizhan Uluduz and Minjung Kim et al.
Evaluation of PVC-Type Insulation Foam Material for Cryogenic Applications
Reprinted from: *Polymers* **2023**, *15*, 1401, doi:10.3390/polym15061401 56

Arif Muhammad, Mkhululi Ncube, Nithish Aravinth and Jacob Muthu
Controlled Deposition of Single-Walled Carbon Nanotubes Doped Nanofibers Mats for Improving the Interlaminar Properties of Glass Fiber Hybrid Composites
Reprinted from: *Polymers* **2023**, *15*, 957, doi:10.3390/polym15040957 76

Hana Jung, Kwak Jin Bae, Yuna Oh, Jeong-Un Jin, Nam-Ho You and Jaesang Yu
Effects on the Thermo-Mechanical and Interfacial Performance of Newly Developed PI-Sized Carbon Fiber–Polyether Ether Ketone Composites: Experiments and Molecular Dynamics Simulations
Reprinted from: *Polymers* **2023**, *15*, 1646, doi:10.3390/polym15071646 90

Oussama Elalaoui
Effect of Short Fibers on Fracture Properties of Epoxy-Based Polymer Concrete Exposed to High Temperatures
Reprinted from: *Polymers* **2023**, *15*, 1078, doi:10.3390/polym15051078 107

Wenyan Lv, Jun Lv, Cunbing Zhu, Ye Zhang, Yongli Cheng and Linghong Zeng et al.
Thermal Stabilities and Flame Retardancy of Polyamide 66 Prepared by In Situ Loading of Amino-Functionalized Polyphosphazene Microspheres
Reprinted from: *Polymers* **2022**, *15*, 218, doi:10.3390/polym15010218 122

Ali A. Rajhi
Mechanical Characterization of Hybrid Nano-Filled Glass/Epoxy Composites
Reprinted from: *Polymers* **2022**, *14*, 4852, doi:10.3390/polym14224852 133

Branka Mušič, Nataša Knez and Janez Bernard
Flame Retardant Behaviour and Physical-Mechanical Properties of Polymer Synergistic Systems in Rigid Polyurethane Foams
Reprinted from: *Polymers* **2022**, *14*, 4616, doi:10.3390/polym14214616 . . . . **148**

About the Editor

SD Jacob Muthu

SD Jacob Muthu is a distinguished researcher in the field of material science and engineering and is well known for his work in polymer composites and nanocomposites. With an illustrious carrier spanning a number of years, he has contributed significantly to his research field. Throughout his academic journey, Dr. Muthu has published numerous journal papers, conference proceedings, and research reports that have enriched the scientific community.

Preface

With the ever-evolving innovation in nanomaterials, polymer nanocomposites have revolutionized engineering industries and technologies across the globe and unlocked new possibilities and functionality for polymers. The combination of nanomaterials with polymers has promoted a new generation of materials that exhibit exceptional mechanical, thermal, electrical and optical properties. This Reprint is a collection of journal papers focused on the various combination of nanocomposites that analyze the intricate behaviour of nanoparticles to improve the functionalities of polymers in various engineering applications.

All the papers in this Reprint have been subjected to a rigorous peer-review process before being accepted for publication. The published articles will benefit broad audiences, including students, research engineers, and professionals seeking to understand the complexities of polymer nanocomposites. Further, this Reprint will offer a wealth of knowledge and insights into applying nanocomposites in various engineering industries.

SD Jacob Muthu
Editor

Article

Dynamic Impact Properties of Carbon-Fiber-Reinforced Phenolic Composites Containing Microfillers

Ibraheem A. Abdulganiyu *, Oluwasegun E. Adesola, Ikechukwuka N. A. Oguocha and Akindele G. Odeshi

Department of Mechanical Engineering, University of Saskatchewan, Saskatoon, SK S7N 5A9, Canada; qvf897@mail.usask.ca (O.E.A.); iko340@campus.usask.ca (I.N.A.O.); ago145@mail.usask.ca (A.G.O.)
* Correspondence: iaa763@mail.usask.ca

Abstract: The addition of nano- and microfillers to carbon-fiber-reinforced polymers (CFRPs) to improve their static mechanical properties is attracting growing research interest because their introduction does not increase the weight of parts made from CFRPs. However, the current understanding of the high strain rate deformation behaviour of CFRPs containing nanofillers/microfillers is limited. The present study investigated the dynamic impact properties of carbon-fiber-reinforced phenolic composites (CFRPCs) modified with microfillers. The CFRPCs were fabricated using 2D woven carbon fibers, two phenolic resole resins (HRJ-15881 and SP-6877), and two microfillers (colloidal silica and silicon carbide (SiC)). The amount of microfillers incorporated into the CFRPCs varied from 0.0 wt.% to 2.0 wt.%. A split-Hopkinson pressure bar (SHPB), operated at momentums of 15 kg m/s and 28 kg m/s, was used to determine the impact properties of the composites. The evolution of damage in the impacted specimens was studied using optical stereomicroscope and scanning electron microscope. It was found that, at an impact momentum of 15 kg m/s, the impact properties of HRJ-15881-based CFRPCs increased with SiC addition up to 1.5 wt.%, while those of SP-6877-based composites increased only up to 0.5 wt.%. At 28 kg m/s, the impact properties of the composites increased up to 0.5 wt.% SiC addition for both SP-6877 and HRJ-15881 based composites. However, the addition of colloidal silica did not improve the dynamic impact properties of composites based on both phenolic resins at both impact momentums. The improvement in the impact properties of composites made with SiC microfiller can be attributed to improvement in crystallinity offered by the α-SiC type microfiller used in this study. No fracture was observed in specimens impacted at an impact momentum of 15 kg m/s. However, at 28 kg m/s, edge chip-off and cracks extending through the surface were observed at lower microfiller addition (≤1 wt.%), which became more pronounced at higher microfiller loading (≥1.5 wt.%).

Keywords: particle-reinforced composites; mechanical properties; split-Hopkinson pressure bar (SHPB); impact behaviour; crystallinity; fracture

1. Introduction

Carbon-fiber-reinforced phenolic composites (CFRPCs) are used widely as structural materials in aeronautical and space industries due to their excellent properties such as low density, low thermal expansion, high thermal stability, high strength and stiffness, and excellent resistance to creep, fatigue, and corrosion [1,2]. Moreover, they can be turned into carbon-carbon composites by converting the phenolic matrix into amorphous carbon through pyrolysis [3–6].

Phenolic resins have attractive properties which make them good matrix materials for fabricating carbon-fiber-reinforced composites. Compared to other polymers, phenolic resins have lower flammability and smoke emission, superior mechanical strength, higher hardness, better heat resistance, chemical resistance, and dimensional stability [7–9]. Phenolic resin material systems have been used for ballistic protective body armor since the 1960s. Phenolic resin blended with polyvinyl butyral (PVB) resin was one of the earliest matrix materials approved for ballistic protective body armor [10]. In the early 1960s, DeBell & Richardson Inc. (Enfield, CT, USA) developed a PVB-phenolic resin system initially for nylon helmet liners. Aircraft structures experience impact damage as a result of bird strikes, turbine blade failures, and inspiration of runaway debris into the engine, with velocities of 100–150 m/s and impact energies of over 100 J [11,12].

The structural performance of polymers and fiber-reinforced polymer composites can be enhanced through the addition of filler materials such as carbon nanotube (CNT), graphene, graphene oxide (GO), nanoclay, non-metallic oxides and carbides, and metallic oxides, carbides, borides, silicates, and silicides. The modification of matrices of CFRP composites with nano- and microfillers has been reported to lead to considerable improvement in their ablation resistance [13–19], interlaminar shear strength (ILSS) [20–27], interlaminar fracture toughness (ILFT) [22,28–30], wear and tribological properties [31–33], as well as flexural properties (strength and elastic modulus) [34,35]. Xu et al. [13] studied the ablation resistance and mechanism of CF-fabric-reinforced phenolic resin composite modified with particles of tantalum disilicide ($TaSi_2$) and reported significant improvement in the anti-ablation property of $TaSi_2$-modified composite in comparison with the unmodified composite. At 50 wt.% $TaSi_2$ addition, the linear ablation rate and mass ablation rate reduced by 12% and 30%, respectively. In a different study, Wang et al. [14] found that the linear ablation rate of SiC particle-modified CF-reinforced phenolic resin composite decreased with increasing SiC content up to 5 wt.%, after which it started to increase. The effect of halloysite nanoclay addition on the mechanical and thermal properties of carbon/glass-fiber-reinforced epoxy resin composites was investigated by Nagaraja et al. [20]. They found that the interlaminar shear strength (ILSS) of nanoclay-modified composites was higher than that of the unmodified composites. At 3 wt.% halloysite nanoclay content, the ILSS of the composites improved by roughly 18% over that of unmodified composites. Lyu et al. [27] reported roughly 55% and 94% increases in ILSS and flexural strength, respectively, for CF/polyetheretherketone (PEEK) composites modified with multi-walled carbon nanotube (MWCNT) when compared to unmodified CF/PEEK composites. Kostagiannakopoulou et al. [29] studied the effect of adding graphene nanoplatelets with different geometrical characteristics on the interlaminar fracture toughness (ILFT) of CFRP composites. They observed that composites containing graphene nanoplatelets with high values of aspect ratio (AR) and specific surface area (SSA) showed higher Mode I and II ILFT than the unmodified composites.

A main challenge arising from the inclusion of micro- and nanofillers in polymer-based composites is that they agglomerate during dispersion in the matrix as a result of their large SSA and van der Waals forces [36–38]. The net effect of agglomeration is that it negatively influences the end properties of the resulting composites. Moreover, pristine carbon fiber is smooth, chemically inert, hydrophobic, and possesses low surface energy. As such, bonding between it and the polymer matrix occurs mostly through the weak van der Waals forces. Some of the methods employed in dispersing fillers in polymer matrices are vigorous mixing, shear mixing, mechanical mixing (e.g., ball-milling and 3-roll milling or calendaring), ultrasonication, magnetic stirring, surface coating of fillers, and chemical and physical surface functionalization [29,36–41]. While a perfect dispersion of the filler in different polymer matrices could be very challenging and difficult to achieve, factors such as the type of the polymer matrix (thermoplastic or thermoset), dimensions and amount of microfillers, the availability of techniques and fabrication processes are considered when selecting a proper technique for filler dispersion [29,41]. The dispersion state of fillers in matrices of polymer composites plays an important role in determining their overall performance. Its evaluation can provide an insight into the relationship between the microstructure and measured properties of the composites [42,43]. The experimental techniques commonly used for evaluating dispersion state include optical microscopy, dynamic light scattering (DLS), ultraviolet-visible-infrared (UV-Vis-IR) spectroscopy, scanning electron microscopy (SEM), transmission electron microscopy (TEM), X-ray diffraction, X-ray microtomography, and small angle scattering techniques, such small angle X-ray scattering (SAXS), small angle light scattering (SALS), and small angle neutron scattering (SANS) [42–46].

A survey of the available literature shows that, whereas several research works (Zhao et al. [47], Mohsin et al. [48], Rouf et al. [49], Lomakin et al. [50], Rouf et al. [51], Li et al. [52])have been carried out on the dynamic mechanical properties of plain CF-reinforced polymer composites, there is little information on the dynamic impact behaviour of microfiller-modified CFRPCs (Chihi et al. [53]) investigated using a split-Hopkinson pressure bar (SHPB) apparatus. Zhao et al. [47] used an electromagnetic SHPB to study the influence of strain rate and loading direction on the compressive failure behaviour of 2D triaxially braided CF fabric/epoxy resin composites at dynamic strain rates of 200 s^{-1} and 220 s^{-1}. The results of the SHPB tests were compared with those of quasi-static tests performed at strain rates of 10^{-4} s^{-1} and 10^{-2} s^{-1}. Both axial and transverse compressive strengths were found to increase with strain rate, which they attributed to change in the mechanisms of damage evolution. Rouf et al. [49] studied the effect of strain rate on the compressive failure characteristics of a 1D non-crimp fabric (1D-NCF) carbon fiber/epoxy composite at strain rates of 0.003/s, 0.2/s, and 325/s. The high strain rate (325/s) test was performed using a SHPB apparatus. A strong strain rate dependency of the compressive strength of the composites was reported. On the other hand, Li et al. [52] studied the strain rate dependency of mechanical properties of warp-knitted and plain weave CF fabric/epoxy resin composites. A quasi-static test was performed at 0.5/s, while dynamic compression tests were performed with a SHPB instrument at strain rates that ranged from roughly 221/s to 1337/s. A weak strain rate dependency of the dynamic compressive strength was reported. Chihi et al. [53] investigated the dynamic response and damage characteristics of CF/epoxy resin composites containing different weight fractions of CNTs (0%, 0.5%, and 2%) using the SHPB equipment. Compression SHPB tests were conducted at impact pressures ranging from 1.4 bars to 2 bars. The results showed that the addition of CNTs significantly improved the mechanical performance of CF/epoxy composites. Within the range of CNT content and impact pressure tested, the maximum compressive strength obtained increased with the impact pressure and weight fraction of CNT.

The goal of this research is to improve the impact response of carbon-fiber-reinforced phenolic composites with microfillers (colloidal silica and micron-sized silicon carbide) at varying amounts and establish a performance threshold limit of the CFRPCs with the microfiller addition. This would help usher in a new class of high-performing CFRP composites for use in high strain rate applications without suffering catastrophic failure in service. In the long run, this research is intended to develop numerical models that could predict the CFRP material performance at high strain rate loading conditions. To develop this numerical model, material data generated at high strain rates are necessary. For this purpose, the SHPB device was used in this work to obtain high strain rates.

2. Materials and Methodology
2.1. Materials

Figure 1 shows the flow chart used in manufacturing CFRP composites containing microfillers. A 3K, 2 × 2 twill weave (PAN) carbon fiber (CF) fabric manufactured by Fibre Glast Developments Corporation (Brookville, OH, USA), with each fabric layer measuring 0.305 mm thick, was used as the carbon fiber material. Two resole-type phenolic resins (HRJ-15881 and SP-6877) procured from SI Group (Schenectady, NY, USA) were used as matrix materials. Their compositions and properties, as provided by the manufacturer, are shown in Table 1. Two microfillers were used to modify the matrices, namely: colloidal silica (406), which was procured from West System Inc., (Bay City, MI, USA), and silicon carbide (SiC) produced by Washington Mills, (North Grafton, MA, USA). The colloidal silica (CS) comprises sand and quartz particles with an average particle size of 0.2–0.3 μm and a density of 50 g/L. The density of the SiC was 3.19 g/cm^3 and its particle size range was 0.3–2.2 μm. Polyethylene glycol (PEG) (product name: Carbowax^{TM} PEG-400), with a density of 1.13 g/mL, which was procured from Fisher Chemical Canada, was used to disperse the colloidal silica in the phenolic resins. PEG has long molecular chains which tend to disrupt the mutual attraction between individual microfiller particles, thereby promoting their uniform dispersion in the phenolic resins [37,38].

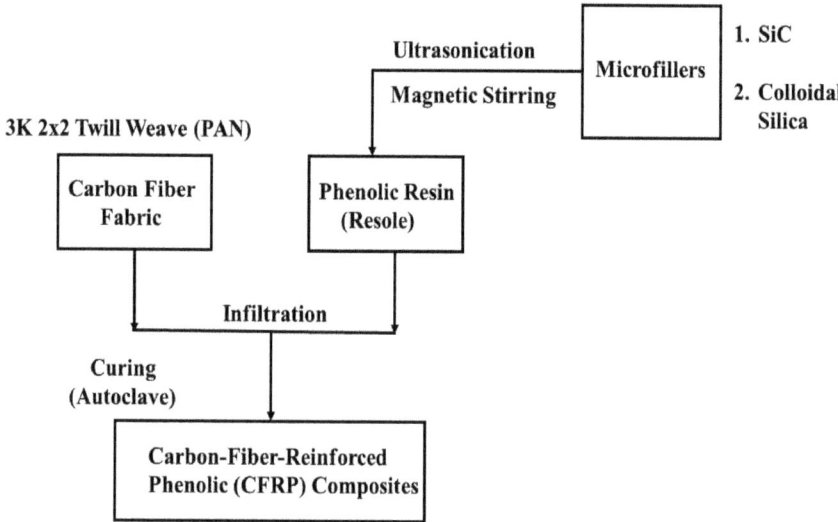

Figure 1. A flow chart showing the procedure for fabricating microfiller-modified CFRP composites.

Table 1. Compositions and properties of the two resole phenolic resins (HRJ-15881 and SP-6877).

Characteristics	HRJ-15881	SP-6877
Solids (%)	76.09	76.09
Phenol (%)	13.61	13.61
pH	8.1	8.1
Viscosity Brookfield (cP)	906	53.1
Gel Time (min.)	12.5	12.5
Formaldehyde (HCHO, (%))	0.5	1.3

2.1.1. Dispersion of Microfillers in the Resole Phenolic Resins

A detailed description of the dispersion of colloidal silica (CS) and SiC microfillers in the phenolic resins can be found in the work of Abdulganiyu et al. [54]. To provide further context, CS was mixed with PEG in a ratio of 2:1 to form a shear-thickening fluid (STF). By shear thickening, we mean an increase in viscosity with increasing rate of shear strain during mechanical deformation [55]. The PEG and CS mixture was stirred using a magnetic stirrer at 1200 rpm for 3 h at room temperature. Afterward, the resulting STF was diluted with ethyl alcohol using a ratio of 3:2 and held for 2 h. The mixture was then added to the phenolic resins in the appropriate amount to obtain 0.5, 1.0, 1.5, and 2.0 wt.% of CS in the resins. To promote CS dispersal in the phenolic resins, the mixture was stirred magnetically at 1200 rpm for 1 h, followed by ultrasonication for 1 h using a Branson 1510 ultrasonicator. The ultrasonication was performed in steps of 10 min with 5 min of rest in between to prevent overheating.

Firstly, SiC particles were weighed in the appropriate proportions to correspond to 0.5, 1.0, 1.5, and 2.0 wt.%. Then, they were mixed with 5 mL ethanol absolute to produce composites containing 0.5 wt.% SiC. The volume of ethanol mixed with SiC particles was increased to 6, 7, and 8 mL, for composites containing 1.0, 1.5, and 2.0 wt.% SiC particles, respectively. The main reason for using ethanol absolute was to help disperse SiC particles in the phenolic resin. In addition, it was anticipated that the ethanol would evaporate when the resole resin is heated during the curing process, leaving well-dispersed SiC particles in the phenol resin matrix of the resulting CFRP composite. The SiC particles-ethanol mixture was then ultrasonicated for a total of 30 min using the Branson 1510 ultrasonicator. The sonication was also carried out in time steps of 10 min with 5 min rest in between to prevent overheating. The ultrasonicated mixture of ethanol and SiC microfillers was then mixed with the phenolic resin and the new mixture was further sonicated for 1 h (in time steps of 10 min with 5 min rest in between to prevent overheating). Finally, after the ultrasonication, the mixture was magnetically stirred at 1200 rpm for 1 h.

2.1.2. Fabrication of CFRP Composites

An in-house mold measuring 170 mm × 60 mm × 10 mm ($l \times w \times t$) was used to manufacture CFRP composite specimens used for dynamic impact testing. The carbon fiber fabric was cut into rectangular sheets measuring 160 mm × 60 mm ($l \times w$). Since the thickness of the internal cavity of the mold is 10 mm and thickness of the carbon fiber fabric sheet is 0.305 mm, the number of reinforcement sheets used (M) was calculated using Equation (1).

$$M = \frac{thickness\ of\ mold\ internal\ cavity}{thickness\ of\ carbon\ fiber\ fabric\ sheet} = \frac{10\ mm}{0.305\ mm} \tag{1}$$

The internal cavity of the mold was sprayed with a SLIDE® Epoxease mold release agent (No. 40614N) produced by Infotrac (USA) and allowed to dry for 25 min. The lubricant made the removal the CFRPCs from the mold after fabrication easy. Using the hand lay-up technique, 33 sheets of the carbon fiber fabric were pressed together to fit into the 10 mm thickness of the internal cavity of the mold. Afterward, the previously prepared mixture of microfillers and phenolic resin was poured into the mold cavity and it slowly impregnated the carbon fiber fabric sheets with the help of gravity.

Curing of the resin-impregnated carbon fiber fabric was carried out in a Parr 4848 pressure reactor autoclave under an argon gas atmosphere at a pressure of 50 bar. The temperature of the autoclave was ramped up from room temperature to 120 °C at a heating rate of 2 °C/min and maintained at 120 °C for 1 h, after which it was cooled to room temperature. The argon gas atmosphere helped to remove gaseous by-products of polymerization reactions that occurred during curing. It also helped to reduce the porosity of the CFRPCs.

2.2. Characterization of Fabricated Materials

2.2.1. X-ray Diffraction (XRD)

Specimens used for XRD experiment were prepared by dipping CFRP composites in liquid nitrogen and crushing them in a clean ceramic crucible to fine aggregates. XRD analysis was obtained with a Rigaku Ultima IV X-Ray Diffractometer (Rigaku Americas Corporation, The Woodlands, TX, USA) using Cu Kα source (λ = 1.54056 Å), tube voltage of 40 kV and tube current of 44 mA. Experimental data were collected using a Multipurpose Attachment, with para-focusing mode. A nickel K$_\beta$ filter was placed at the receiving end. X-ray diffractograms were acquired in 2θ from 5° to 60° at a scan rate of 0.5°/min and a step size of 0.02. The software used for background correction and data smoothening was PANalytical X'Pert Highscore, while Origin 2019 (OriginLab, Northampton, MA, USA) software was used for plotting and analyzing the obtained XRD data.

2.2.2. Dynamic Impact Test

The high strain rate behaviour of fabricated CFRP composites was investigated using a split-Hopkinson pressure bar (SHPB) apparatus, the schematic diagram of which is presented in Figure 2. As can be seen, it comprises three bars, namely: striker, incident, and transmitter bars. The transmitter and incident bars were machined from aluminum alloy 7075-T651 and have a diameter of 38 mm. Figure 3 shows a photographic image of typical impact test specimens, each measuring 12 mm × 12 mm × 10 mm. The surfaces of test specimens in contact with the incident and the transmitted bars were lubricated with Vaseline to reduce friction between the contacting end surfaces. With the specimen placed along the thickness between the incident and transmitted bars, the striker bar was fired with the aid of the gas gun toward the incident bar at pressures of 50 kPa and 80 kPa which corresponded to impact momentums of 15 kg m/s and 28 kg m/s, respectively. Five specimens of each CFRP composite and neat cured resin were tested at each impact momentum.

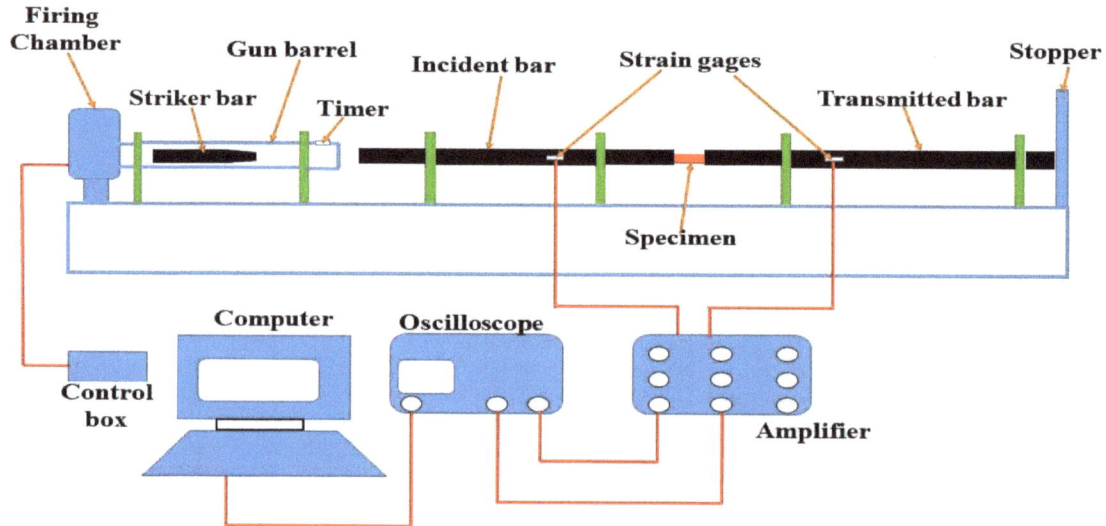

Figure 2. A schematic diagram of the split-Hopkinson pressure bar (SHPB) system.

Figure 3. Photographic image showing typical dynamic compression test specimens.

During SHPB tests, an incident stress pulse is produced when the striker bar impacts the incident bar which is propagated along the incident bar toward the specimen–incident bar interface. A portion of this stress pulse is back-reflected through the incident bar, while the remainder is propagated to the transmitter bar through the specimen. Since a constant strain rate helps to eliminate the radial inertia effects on test specimens during SHPB tests [56], a pulse shaper with a thickness of 12 mm but the same diameter as the incident and the transmitter bars was attached to the end of the incident bar. As such, on firing at a given impact momentum, the striker came into contact with the pulse shaper to help achieve a constant strain rate. Strain gauges were used to capture the generated elastic strain waves (incident, reflected, and transmitted) traveling through the incident and transmitter bars. Amplification and conditioning of the elastic strain waves were carried out using the connected strain conditioner-amplifier system. The output from the conditioner-amplifier system was linked to the mixed-signal digital oscilloscope which, in turn, was connected to the computer. The oscilloscope output was controlled and graphically displayed by in-house developed LabView software program. Based on stress equilibrium and obtained experimental data, the engineering or nominal stress (σ_e), engineering strain (ε_e), and initial engineering strain rate ($\dot{\varepsilon}_e$) were calculated using Equations (2)–(4), respectively [57]:

$$\sigma_e = \frac{A_b}{A_s} E_b \varepsilon_T \qquad (2)$$

$$\varepsilon_e = -2 \frac{C_b}{L_s} \int_0^t \varepsilon_R dt \qquad (3)$$

$$\dot{\varepsilon}_e = -2 \frac{C_b}{L_s} \varepsilon_R \qquad (4)$$

where A_S and A_b stand for cross-sectional areas of the specimen and bars, respectively; ε_T and ε_R stand for transmitted and reflected strain waves, respectively; C_b is the elastic wave velocity in the bars; E_B stands for the Young's modulus of the bar materials; while L_s and t represent the original length of the specimen and deformation time, respectively. True stress (σ_t) vs. true strain (ε_t) curves were generated based on Equations (5) and (6), while the true strain rate ($\dot{\varepsilon}_t$) was calculated using Equation (7) [58].

$$\sigma_t = \sigma_e(1 - \varepsilon_e) \qquad (5)$$

$$\varepsilon_t = -ln(1 - \varepsilon_e) \qquad (6)$$

$$\dot{\varepsilon}_t = \frac{\dot{\varepsilon}_e}{(1 - \varepsilon_e)} \qquad (7)$$

The two firing pressures used in this study, 50 kPa and 80 kPa, generated average true strain rates of 880 s^{-1} and 2150 s^{-1} in the test specimens, respectively.

2.2.3. Microstructure Examination

Specimens with dimensions 10 mm × 10 mm × 10 mm, cut from the fabricated rectangular plates of CFRP composites using a Buehler abrasive cutter (Model 95, C1800), were used for microstructure analysis. They were ground initially using 320, 500, 800, and 1200 SiC grit emery papers, followed by fine grinding using 2000 and 4000 SiC grit emery papers and finally polished using a 5 µm MD-Dac cloth with 5 µm MD-Dac suspension and a 1 µm MD-Nap cloth with 1 µm MD-Nap suspension. Microstructures of the polished and fractured specimens were analyzed using optical microscopy (OM, Nikon Optihot stereomicroscope, Melville, NY, USA, interfaced with a PAX-it digital camera) and scanning electron microscopy (SEM, JEOL-JSM-6010LV, Peabody, MA, USA). To reduce charging and improve the image quality, SEM specimens were gold-coated using an Edward S150B sputter coater (BOC Edwards, Crawley, UK). The SEM was operated at an accelerating voltage of 20 kV and images were acquired in the secondary electron (SE) imaging mode.

3. Results and Discussion
3.1. XRD

To properly index the X-ray diffraction (XRD) patterns obtained experimentally for microfillers and carbon fiber (CF), they were compared to standard diffraction peaks. Indexing of experimental diffraction peaks of the CF was carried out using the Mercury software (version 4.3.0) and graphite crystallographic information (.cif) file (deposition number: 918549). The graphite cif data were sourced from the crystal structure database of the Cambridge crystallographic data center (CCDC). The simulation of theoretical diffraction patterns of graphite was performed using full width at half-maximum (FWHM) values of 1.0, 2.0, 3.0, 4.0, and 5.0. The diffraction patterns generated using these FWHM values are shown in Figure 4. Clearly, the characteristic intense peak at roughly $2\theta = 26.2°$ (002) is followed by minor peaks at $2\theta \sim 44.0°$ (101) and $2\theta \sim 54.0°$ (004). The existence of additional peaks at $2\theta \sim 42.3°$ (100) and $2\theta \sim 50.5°$ (102) depends on the FWHM value used.

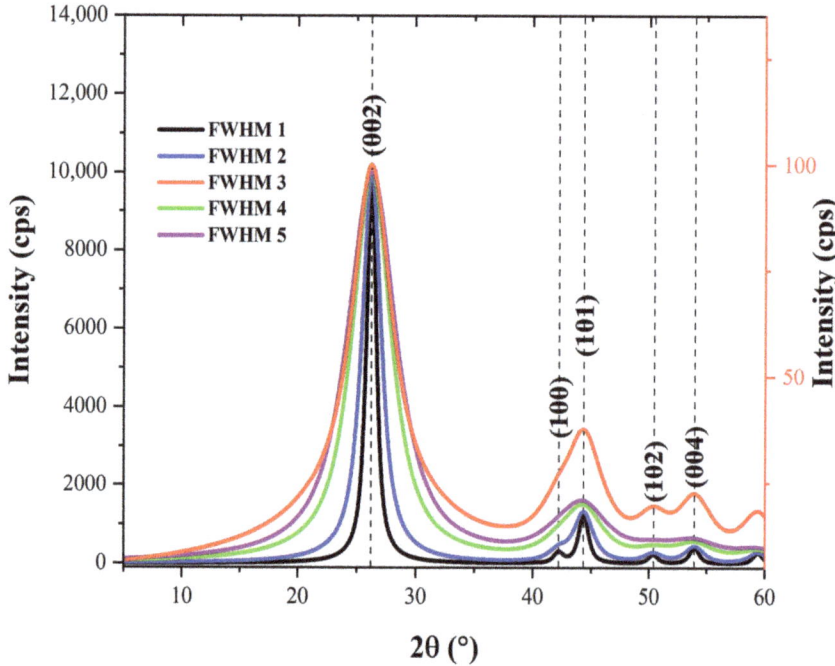

Figure 4. Diffraction patterns of graphite computed using different values of FWHM.

The calculated diffraction patterns were compared with the diffraction pattern obtained experimentally for the carbon fiber used in the present study (designated as Exp. CF). It was found that the XRD pattern calculated using an FWHM value of 4.0 matched the XRD pattern obtained experimentally for CF (Exp. CF). The two XRD patterns are co-plotted in Figure 5 for comparison. For Exp. CF, it can be seen that the characteristic peaks which appear at $2\theta \sim 25.7°$ (002), $2\theta \sim 44.1°$ (101), and $2\theta \sim 54°$ (004) are consistent with those obtained for the XRD pattern calculated using an FWHM value of 4.0. Nevertheless, a close look at the two diffraction patterns shows that the calculated (002) peak differs by 0.5° from that of Ex. CF, while the difference is 0.1° for the (101) peak.

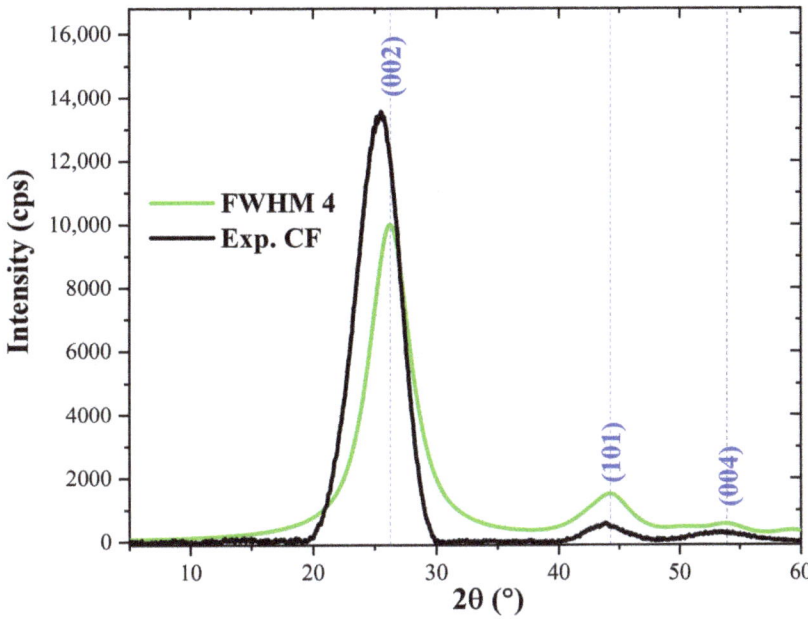

Figure 5. Diffraction pattern of graphite computed using an FWHM value of 4.0 compared with the experimental diffraction pattern obtained for carbon fiber (Exp. CF).

SiC exists in different polymorphs with α-SiC (6H hexagonal) and β-SiC (3C-cubic) being the most stable structures [59,60]. While α-SiC possesses a hexagonal crystal structure and is made at temperature higher than 1700 °C, β-SiC has a cubic crystal structure and is prepared at temperatures lower than 1700 °C. Moissanite occurs naturally as SiC with the 6H hexagonal polymorph [61]. As such, it was used as a standard material with which to compare the SiC microfiller used in this study. The cif file of moissanite used in the present study was retrieved from the American mineralogy crystal structure database (AMCSD). The Mercury software was used to index the diffraction pattern of moissanite generated with an FWHM value of 0.6. The reason why 0.6 was used was because, as shown in Figure 6, the resulting diffraction pattern looked closest to that of the SiC microfiller (designated as Exp. SiC) used in modifying the matrix of CFRP composites developed in the present study. The characteristic features of the diffraction pattern of moissanite (α–SiC) (black colour) are diffraction peaks at 2θ~34.2° (101), 35.7° (102), 38.3° (103), 41.4° (104), 45.3° (105), 54.7° (107), and 60.9° (108). Several of these peaks appeared in the diffraction pattern of Exp. SiC (red colour), indicating that the SiC microfiller used is of the α–SiC type.

For colloidal silica (CS), its cif file could not be found to generate the calculated XRD pattern with which to compare the diffraction pattern of the colloidal silica used in this study (Exp. CS). However, for the Exp. CS, the indexing of the peaks was obtained and compared with values from the literature. Munasir et al. [62] and Jeon et al. [63] obtained (101) peaks for colloidal silica at 2θ~21.5° using the International Center for Diffraction Data (ICDD) PDF-2 database (PDF-2 No. 01-087-2096) and Inorganic Crystal Structure Database (ICSD) as standards. In the present study, a prominent peak at 2θ~21.76° was obtained for the colloidal silica corresponding to the (101) peak from the literature. We can infer that these values agree with one another, even though there is a 0.2° difference between the values reported in the literature and those obtained in this study. The corresponding XRD pattern of the colloidal silica microfiller is shown in corresponding XRD diffractograms, in relation to when colloidal silica microfiller is used.

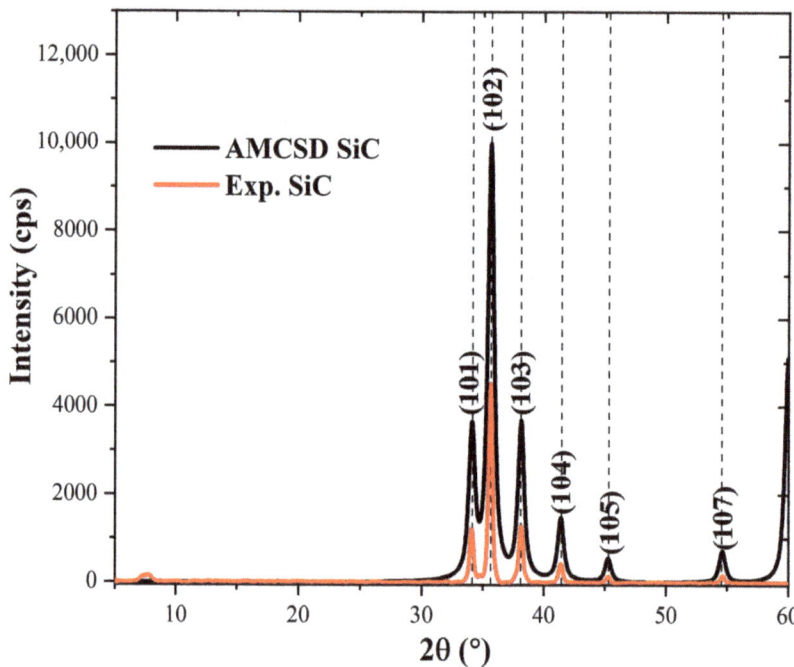

Figure 6. Calculated XRD pattern of α-SiC compared with that of the SiC microfiller (Exp. SiC).

A comparison of XRD patterns obtained for CFRPCs made from phenolic resins HRJ-15881 and SP-6877 with those obtained for CF and monolithic resin is presented in Figure 7. Two conspicuous peaks at 2θ~18.8° and 25.7° for HRJ-15881 resin, and 2θ~18.6° and 25.7° for SP-6877 resin resulting from the addition of CF to the resins, can be seen. Obviously, this is due to the fact that the identities of the constituent materials were unaffected by the manufacturing process used in this study. Furthermore, it is seen that the intensity of the characteristic (002) peak of SP-6877 resin decreased, probably due to its low viscosity.

Figures 8 and 9 show, respectively, XRD patterns obtained for CFRPCs modified with SiC and CS microfillers for HRJ-15881 and SP-6877 resins. The addition of SiC filler to CFRP composites made with the two phenolic resins caused the intensity of the (002) peak to decrease. Sekhar and Varghese [64] investigated the thermal, mechanical, and rheological properties of phenolic resin modified with intercalated graphite bisulfate and found that the characteristic diffraction peak of the graphite moved to the left and its intensity reduced in comparison to the characteristic peak of natural graphite. Based on these results, it was concluded that the intercalated graphite truly intercalates and contains bisulfate. Also, Ki Park et al. [65] investigated the structural and electrochemical evolution of structured V_2O_5 microspheres during Li^+ intercalation and found progressive changes in the characteristic diffraction peak of V_2O_5 as phase transformation occurred during Li^+ intercalation. The characteristic (001) peak of V_2O_5 shifted left to a lower 2θ angle. The shift was attributed to α-V_2O_5 transformation to $Li_xV_2O_5$ at the onset of Li^+ intercalation. In the present study, the decrease in the intensity of the (002) peak of SiC-modified phenolic resin could be due to the penetration of the phenolic matrix by SiC microfillers. Furthermore, the characteristic (102) peak of α-SiC occurred at 35.7° for CFRPCs made with SiC-modified HRJ-15881 and SP-6877 phenolic resins.

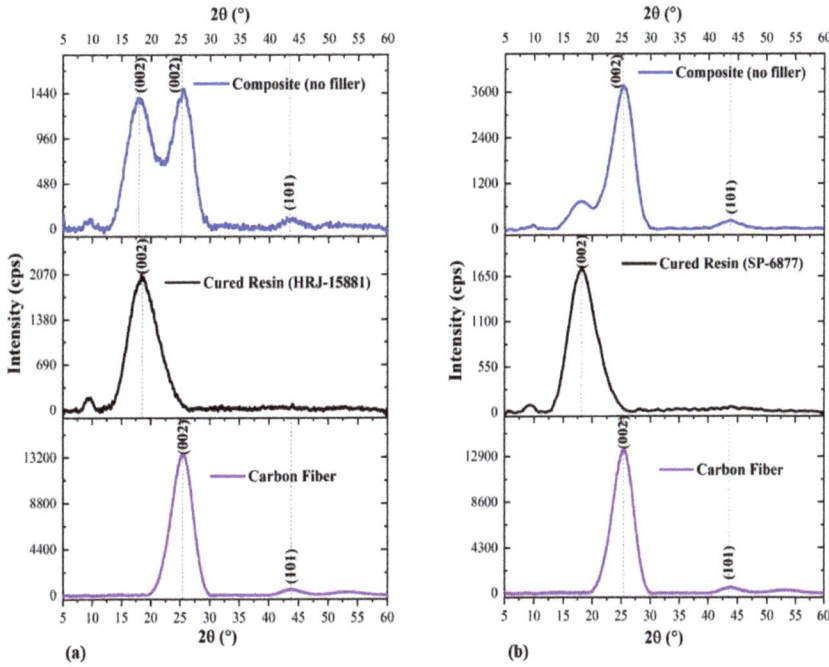

Figure 7. XRD patterns obtained for carbon fiber, monolithic resins, and their composites: (**a**) HRJ-15881 resin and its composite and (**b**) SP-6877 resin and its composite.

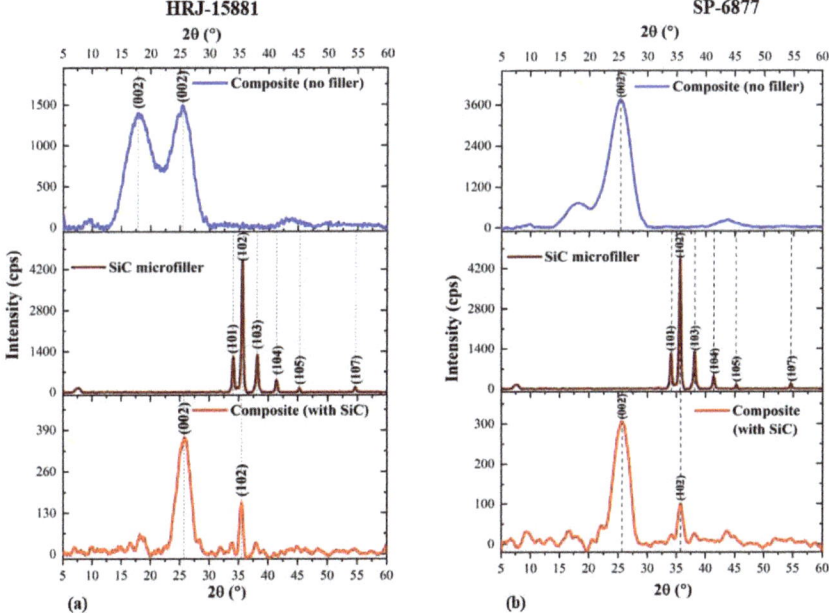

Figure 8. XRD diffraction patterns of CFRP composites modified with 0.5 wt.% SiC microfiller: (**a**) based on HRJ-15881 resin and (**b**) based on SP-6877 resin.

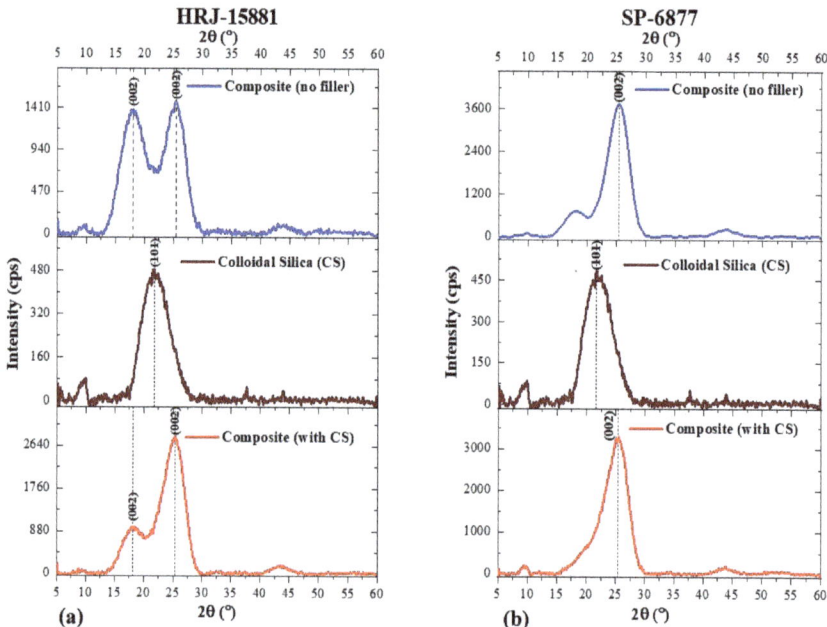

Figure 9. XRD patterns of CFRP composites modified with 0.5 wt.% CS: (**a**) HRJ-15881 resin and (**b**) SP-6877 resin.

Similar to the results presented in Figure 8, it is seen in Figure 9 that CS addition to the CFRP composites caused the intensity of the characteristic (002) peaks of HRJ-15881 and SP-6877 resins to reduce, which is an indication of the presence of CS in the respective matrices of the CF composites manufactured using both resins [64,65].

Figures 10 and 11, respectively, show changes in the intensity and angular position of XRD patterns of neat HRJ-15881 and SP-6877 resins due to the addition of SiC and CS microfillers. Figure 10 shows that SiC microfiller addition (1.0 wt.%) resulted in the existence of a (002) diffraction peak at 2θ~18.4° for both resins. The peaks at 2θ~34.2° (101), 35.7° (102), and 38.3° (103) are characteristic features of α–SiC as shown previously in Figure 8, which confirms that SiC microfillers are present and intercalated in the phenolics [64,65]. Figure 11 shows the effect of adding 1.0 wt.% CS to both resins. The prominent peak at 2θ~18.2° (green colour) shifted slightly to the left when compared with the characteristic peaks of individually cured resins which appeared at 2θ~18.8° and 2θ~18.6° for HRJ-15881 and SP-6877 resins, respectively. The peak shift and decrease in intensity due to CS addition may be due to its intercalation in the phenolic resins. Therefore, on the basis of the XRD patterns shown in Figure 11, it could be concluded that CS fillers are present and intercalated in both phenolic matrices [64,65].

3.2. Dynamic Impact Properties

3.2.1. Impact Properties Obtained at a Momentum of 15 kg m/s

Figures 12–16 present results obtained from SHPB tests carried out at an impact momentum (IM) of 15 kg m/s. Figure 12 shows the true stress–true strain curves obtained for unmodified CFRP composites. Figures 13 and 14 show, respectively, the true stress–true strain and maximum true stress plots obtained for SiC-modified CFRP composites. Similarly, Figures 15 and 16 present the true stress–true strain and maximum true stress plots obtained for CS-modified CFRP composites, respectively.

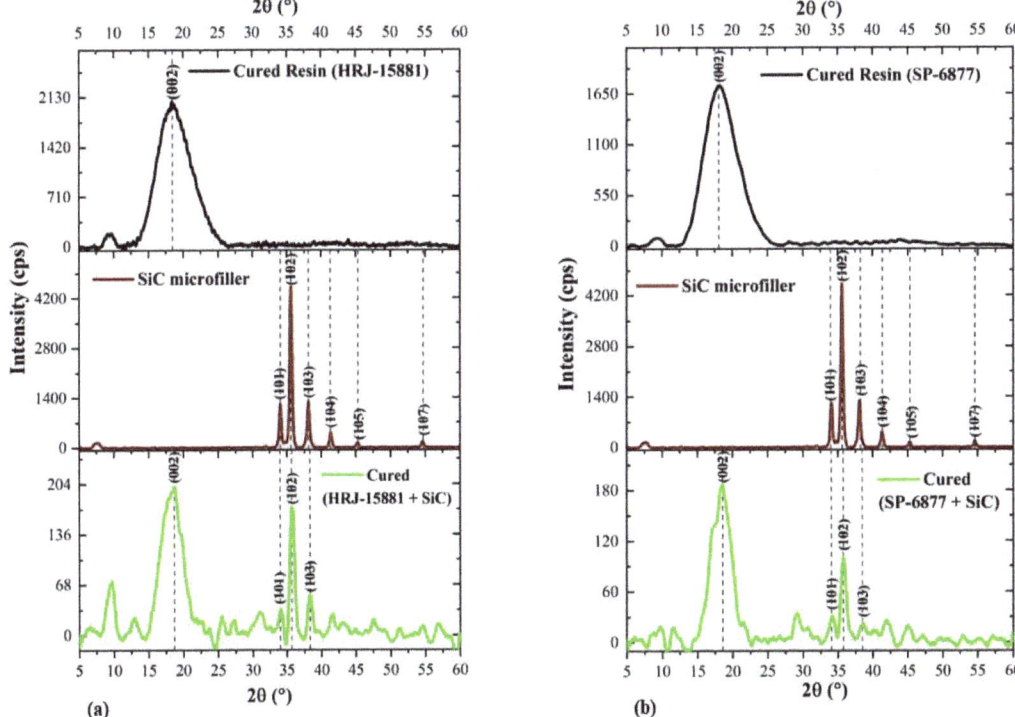

Figure 10. XRD patterns of phenolic resins modified with 1 wt.% SiC microfiller: (a) HRJ-15881 and (b) SP-6877.

As shown in Figure 12, impact strength and toughness of the unmodified CFRP composites fabricated with HRJ-15881 phenolic resin are better than those of unmodified composites made with SP-6877 resin. A close examination of Figure 13 shows that up to 1.5 wt.% SiC addition, the CFRP composites based on HRJ-15881 matrix showed better impact strength than the unmodified CFRP composites. However, the impact strength of composites fabricated with SP-6877 resin improved over those of unmodified composites with 0.5 wt.% SiC addition, beyond which it decreased with increasing SiC content.

A plot of maximum true stresses obtained for CFRP composites modified with SiC microfiller is shown in Figure 14. The observed trend is similar to that of the true stress–true strain curves of Figure 13. It can be seen that the impact strength of CFRP composites based on HRJ-15881 matrix is higher than the impact strengths of those made with unmodified and SiC filler-modified SP-6877 phenolic matrix.

Figure 15 shows that the impact strength of CFRP composites modified with CS filler generally deteriorated with increasing microfiller content except for composites based on HRJ-15881 matrix with 1.5 wt.% and 2.0 wt.% CS. The cause of this discrepancy in impact strength within this range of CS content is not clear. The observed trend of reduction in impact strength with increasing CS content could be attributed to particle agglomeration. Abdulganiyu et al. [54] observed a deterioration in flexural properties with colloidal silica addition to CFRP composites owing to the agglomeration of colloidal silica particles. Nevertheless, a close look at Figure 16 indicates that the impact strength of CFRP composites based on HRJ-15881 matrix is higher than those based on SP-6877 matrix at all CS filler contents.

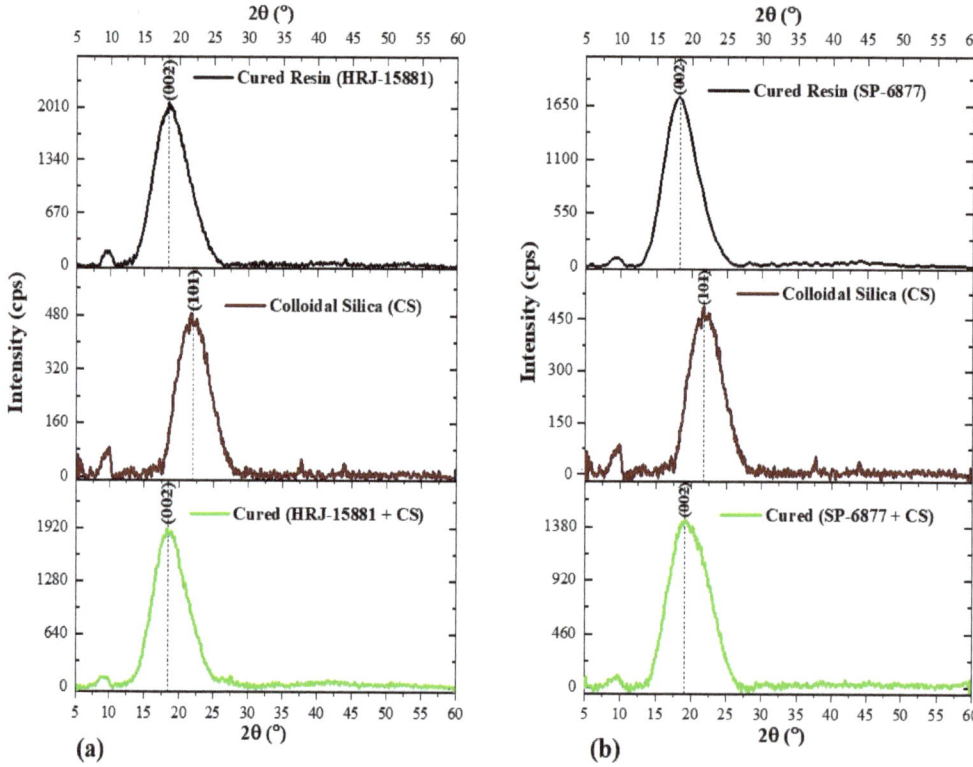

Figure 11. XRD diffraction patterns obtained for phenolic resins modified with 1 wt.% CS: (**a**) HRJ-15881 and (**b**) SP-6877.

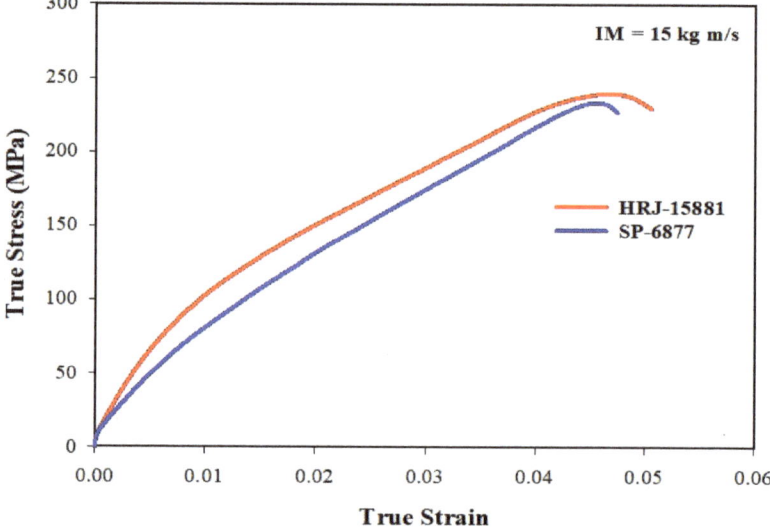

Figure 12. True stress–strain curves of unmodified CFRP composites (IM = 15 kg m/s).

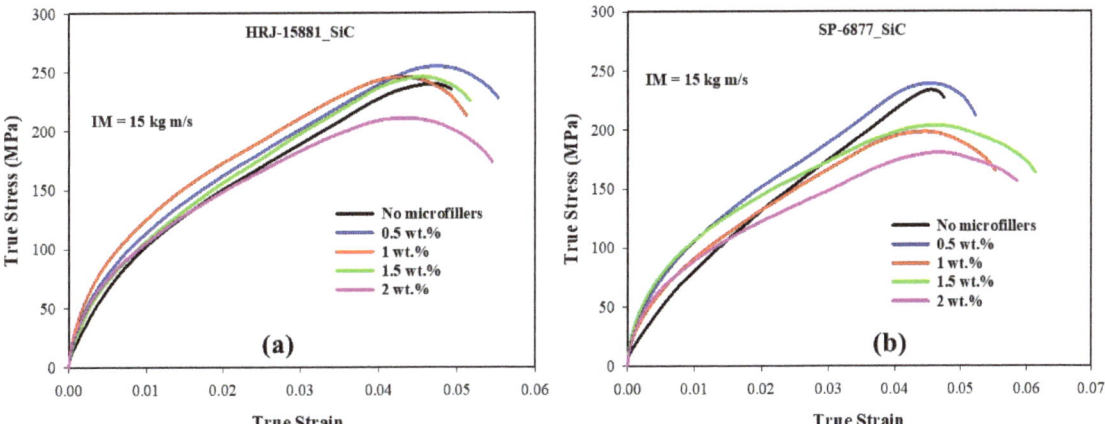

Figure 13. Typical true stress–true strain curves of SiC filler-modified CFRP composites (IM = 15 kg m/s): (**a**) HRJ-15881 and (**b**) SP-6877.

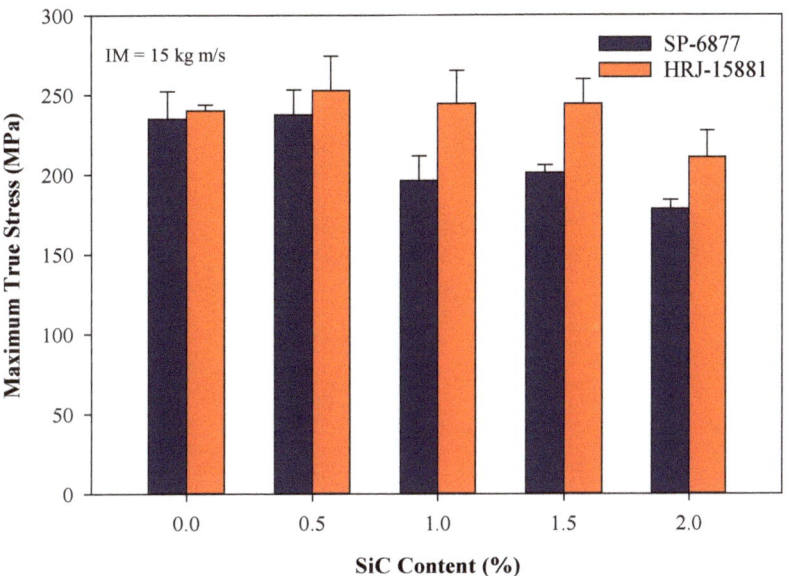

Figure 14. Comparison of maximum true stresses of SiC filler-modified CFRP composites made with the two phenolic matrices (IM = 15 kg m/s). Error bars are based on standard deviation.

3.2.2. Impact Properties Obtained at a Momentum of 28 kg m/s

Figures 17–21 show the results of SHPB tests performed on specimens of CFRP composites at a momentum of 28 kg m/s. Figure 17 shows the true stress vs. true strain curves obtained for unmodified CFRP composites, while Figures 18 and 19 show, respectively, the true stress–true strain curves and maximum true stresses obtained for SiC filler-modified composites. Graphs of true stress vs. true strain and maximum true stresses obtained for CS filler-modified composites are shown, respectively, in Figures 20 and 21.

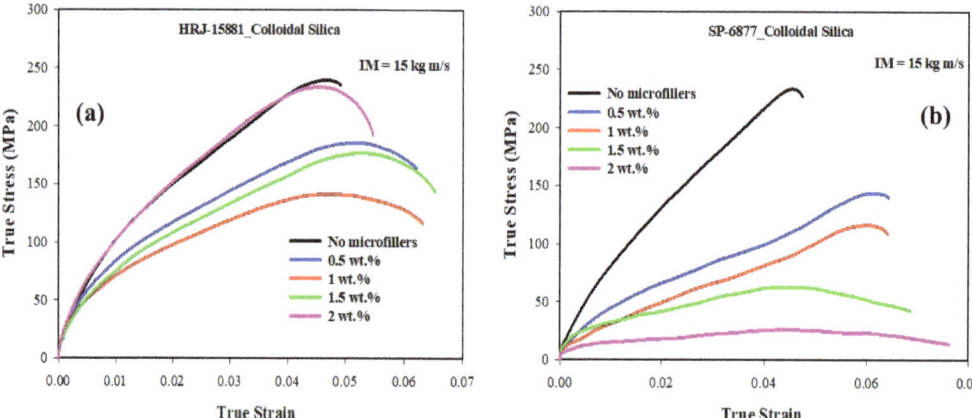

Figure 15. Typical true stress–true strain curves of CS filler-modified CFRP composites (IM = 15 kg m/s): (**a**) HRJ-15881 and (**b**) SP-6877.

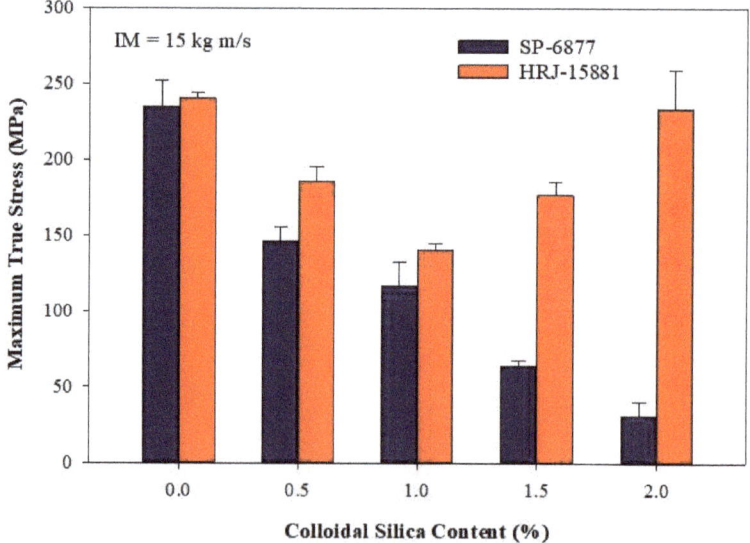

Figure 16. Comparison of maximum true stresses of CS filler-modified CFRP composites made with the two phenolic matrices (IM = 15 kg m/s). Error bars are based on standard deviation.

Figure 17 shows that the impact strength of unmodified CFRP composites based on HRJ-15881 matrix is higher than that based on SP-6877 matrix, which is consistent with the results of unmodified composites tested at 15 kg m/s (see Figure 12). Figures 18 and 19 show that, within the range of filler content tested in this study, the impact strength of SiC filler-modified CFRP composites based on HRJ-15881 matrix is higher than that of those based on SP-6877 resin. A close examination of the true stress–true strain curves in Figure 18 shows that modification of SP-6877 resin with SiC filler improved the impact strength of the resulting composites above that of the unmodified composite at all SiC contents. In contrast, impact strength improvement is seen in HRJ-15881-based composites at only 0.5 wt.% SiC filler addition. Further addition of SiC filler beyond this level resulted in a decrease in impact strength of the resulting composites.

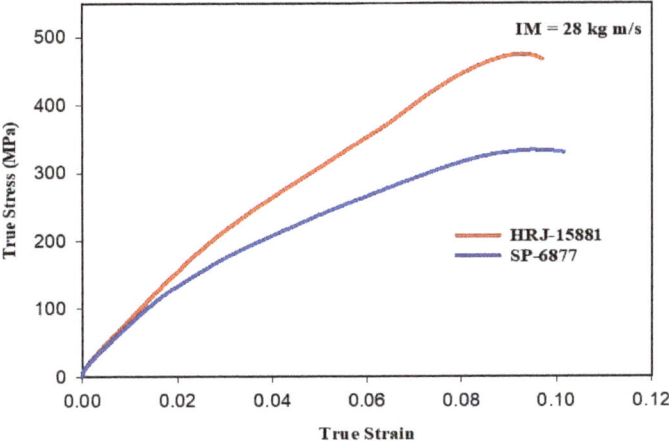

Figure 17. Typical true stress–true strain curves of unmodified CFRP composites (IM = 28 kg m/s).

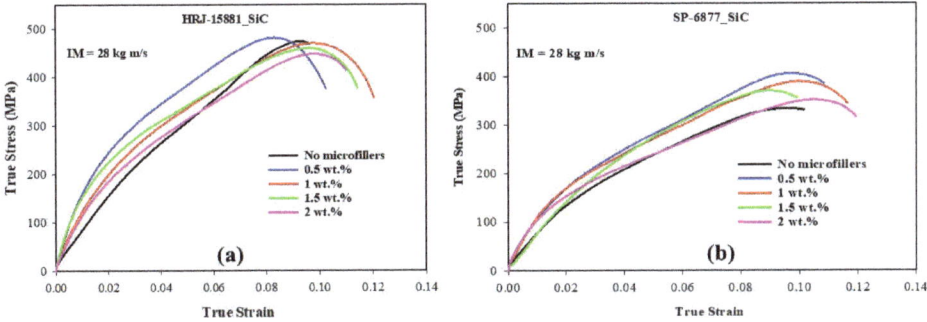

Figure 18. Typical true stress–true strain curves of SiC filler-modified CFRP composites (IM = 28 kg/m/s): (**a**) HRJ-15881 and (**b**) SP-6877.

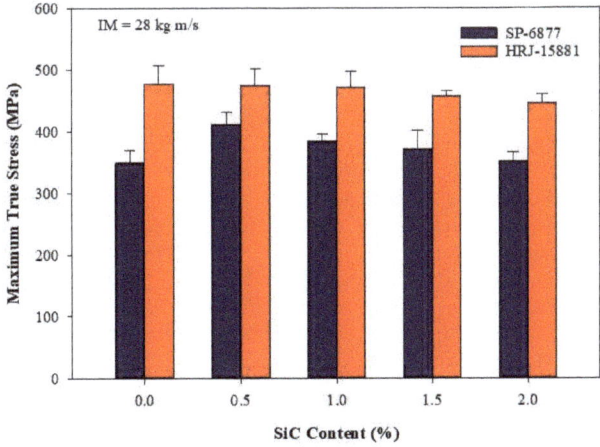

Figure 19. Comparison of maximum true stresses of SiC filler-modified CFRP composites based on the two resins (IM = 28 kg m/s). Error bars are based on standard deviation.

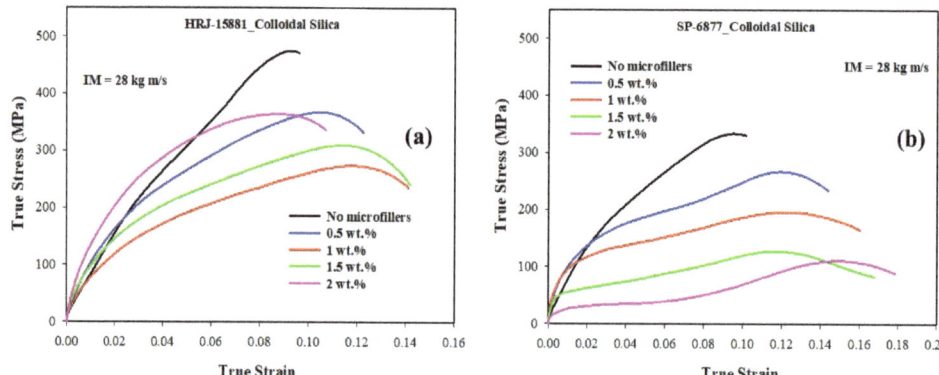

Figure 20. Typical true stress–true strain curves of CFRP composites modified with CS filler (IM = 28 kg m/s): (**a**) HRJ-15881 and (**b**) SP-6877.

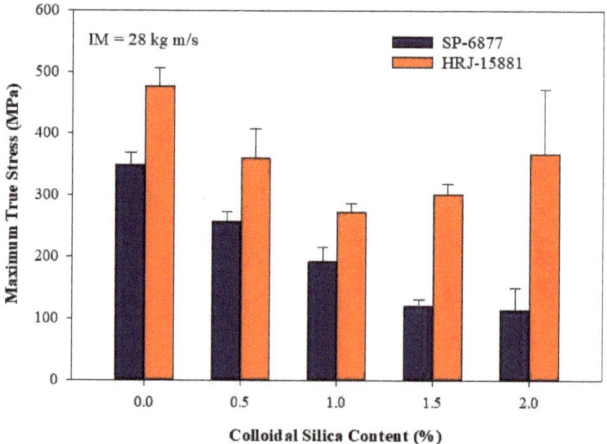

Figure 21. Comparison of maximum true stresses of CS filler-modified CFRP composites based on the two resins (IM = 28 kg m/s). Error bars are based on standard deviation.

As reported previously for composites tested at 15 kg/m/s, Figures 20 and 21 show that modification of both resins with CS filler degraded the impact resistance of the resulting CFRP composites. Nevertheless, within the range of CS content studied, the impact strength of CFRP composites based on HRJ-15881 resin is higher those based on SP-6877 matrix at all CS contents. Furthermore, the impact strength of composites based on SP-6877 matrix generally decreased with increasing CS content. In contrast, the decrease in the impact strength of composites based on HRJ-15881 matrix did not follow a consistent trend. For all intents and purposes, the trend of the impact strength results obtained for composites tested at 15 kg m/s is similar to those tested at 28 kg m/s.

It is evident from the preceding results of SHPB tests, conducted at 15 kg m/s and 28 kg m/s, that the modification of HRJ-15881 and SP-6877 phenolic matrices with SiC filler enhanced the impact resistance of their respective CFRP composites. In contrast, modifying the resins with CS reduced the impact strength of the resulting composites. Broadly, the trends of impact strengths obtained in the present study are consistent with the previous work by Abulganiyu et al. [54] in which the flexural properties of CFRP composites improved with SiC filler addition, but deteriorated with CS addition due to the agglomeration of CS particles.

3.3. Microstructural Examination

Table 2 provides an overview of a visual examination of the failure conditions of specimens of CFRP composites subjected to SHPB testing at the two momentums. At 15 kg m/s, none of the specimens, irrespective of whether they were modified with SiC and CS microfillers or unmodified, fractured. Nevertheless, at 28 kg m/s, specimens of unmodified CFRP composite fractured. Since the fracture features observed for composites based on the two resins were similar, only CFRP composites based on HRJ-15881 matrix would be used as representatives to avoid duplication.

Table 2. Visual inspection results obtained for CFRP composite specimens tested at two impact momentums.

Momentum	Filler Content	Did Fracture Occur?	
		HRJ-15881	SP-6877
15 kg m/s	None	No	No
28 kg m/s		Yes	Yes
15 kg m/s	0.5 wt.% SiC	No	No
	1.0 wt.% SiC	No	No
	1.5 wt.% SiC	No	No
	2.0 wt.% SiC	No	No
28 kg m/s	0.5 wt.% SiC	Yes	Yes
	1.0 wt.% SiC	Yes	Yes
	1.5 wt.% SiC	No	No
	2.0 wt.% SiC	No	No
15 kg m/s	0.5 wt.% CS	No	No
	1.0 wt.% CS	No	No
	1.5 wt.% CS	No	No
	2.0 wt.% CS	No	No
28 kg m/s	0.5 wt.% CS	No	No
	1.0 wt.% CS	No	No
	1.5 wt.% CS	No	No
	2.0 wt.% CS	Yes	Yes

For SiC filler-modified CFRP composites tested at 28 kg m/s, only specimens modified with 0.5 and 1.0 wt.% SiC fractured. Again, as the failure features found in these two CFRP composites were similar, only the detailed microstructures of composites with 0.5 wt.% SiC will be presented. Specimens modified with 1.5 wt.% and 2.0 wt.% SiC filler suffered no fracture due to impact loading at 28 kg m/s. For CS-modified CFRP composites, no fracture was observed below 2.0 wt.% CS content. At 2.0 wt.% CS, specimens of composites based on both resins fractured. Again, failure features were similar for composites based on both resins and, as such, subsequent discussion will focus on HRJ-15881-based composites.

Figures 22–24 show typical optical images of failed specimens of CFRP composites tested at 28 kg m/s. Figure 22 shows typical surfaces of failed specimens of unmodified CFRP composite, Figure 23 shows the same in the context of a CFRP composite modified with 0.5 wt.% SiC, while Figure 24 shows the same regarding a CFRP composite modified with 2.0 wt.% CS. The failure was typified by splitting the specimens into two, with the main failure mode being delamination. In addition, the failed specimens featured undulating, rough surfaces, which indicates that the matrix and fiber bundles ruptured during testing. This type of surface characteristic was more pronounced in SiC filler-modified CFRP composites than in those modified with CS filler. The surface morphology of composites modified with CS filler and those containing no fillers was similar.

Figure 22. Typical optical macrographs of specimens of unmodified CFRP composites, based on HRJ-15881 resin, which were tested at 28 kg m/s.

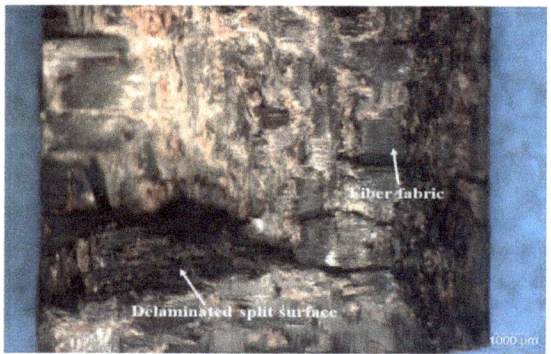

Figure 23. Typical optical macrograph of a specimen of HRJ-15881-based CFRP composite modified with 0.5 wt.% SiC tested at 28 kg m/s.

Figure 24. Typical optical macrograph a specimen of HRJ-15881-based CFRP composite modified with 2 wt.% CS tested at 28 kg m/s.

Figures 25–27 present SEM micrographs of specimens of modified and unmodified CFRP composites that were tested 28 kg m/s. The main failure modes of the unmodified CFRP composites (Figure 25) were CF bundle rupture, specimen splitting, and delamination. Furthermore, in spite of the observed defects, it could be seen from Figure 25 that the CF fabric was impregnated by the phenolic resin quite well. The observed good adhesion between the matrix and CF could explain the reason why the fiber bundles remained intact despite rupture during impact testing.

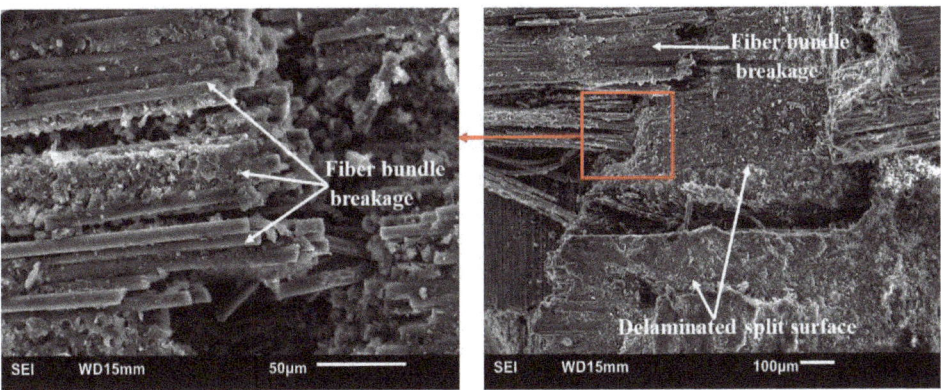

Figure 25. Typical SEM micrographs of unmodified HRJ-15881-based CFRP composites after SHPB testing at 28 kg m/s.

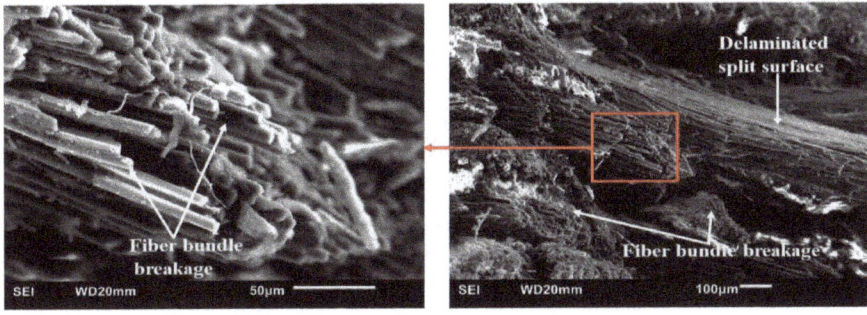

Figure 26. Typical SEM fractographs of HRJ-15881-based CFRP composite modified with 0.5 wt.% SiC after SHPB testing at 28 kg m/s.

Figure 27. Typical SEM fractographs of HRJ-15881-based CFRP composites modified with 2 wt.% CS after SHPB testing at 28 kg m/s.

Similarly, the failure mode of specimens of CFRP composites modified with 0.5 wt.% SiC (Figure 26) is also characterized by fiber bundle breakage and delamination. Nevertheless, a close comparison of Figures 25 and 26 shows that fiber bundle breakage is more pronounced in the SiC filler-modified composites than in the unmodified composites. This may explain the basis for the presence of undulating, jagged ridged surfaces in the optical images of tested specimen shown previously in Figures 22–24. Furthermore, Figure 26 shows there is fairly good bonding between the phenolic matrix and CF.

Figure 27 shows typical SEM micrographs of failed specimens of CFRP composites modified with 2 wt.% CS. As shown previously in Figures 25 and 26, the failure mode is typified by CF bundle breakage and delamination. A matrix crack is shown in the image on the left. Moreover, the micrographs indicate there is good bonding between the matrix and CF, which could be attributed to the good impregnation of CF preforms by the resins.

4. Conclusions

The possibility of enhancing the resistance of CFRP composites to dynamic impact loading by modifying the matrix with SiC and CS microfillers was explored in this study. Plates of CFRP composites were produced using 2D woven CFs, microfillers, and two phenolic resins (SP-6877 and HRJ-15881). SHPB testing of specimens of the composites was conducted at two impact momentums, 15 kg m/s and 28 kg m/s. XRD diffractograms showed that the microfillers were intercalated in the phenolic resins. At 15 kg m/s, HRJ-15881-based CFRP composites modified with 1.5 wt.% SiC gave the highest impact strength. At 28 kg m/s, the impact strength of HRJ-15881-based CFRP composites was higher than that of the SP-6877-based CFRP composites at each level of SiC filler addition. In addition, in the range of SiC content used in this study (0.5–2.0 wt.%), the impact strength of SP-6877-based CFRP composites was better than that of its composite containing no SiC filler. In contrast, only the impact strength of HRJ-15881-based composite modified with 0.5 wt.% SiC was higher than its composite without SiC filler. Nevertheless, the modification of the two resins with CS degraded the impact strength of the resulting composites.

A microstructure examination showed that, at 15 kg m/s, no fracture occurred in the CFRP composites with and without microfillers. However, at 28 kg m/s, the CFRP composites failed by splitting, edge chip off, and crack extension through the matrix, which was more pronounced in the CFRP composites containing colloidal silica microfillers.

Author Contributions: Conceptualization, I.A.A., A.G.O. and I.N.A.O.; Methodology, I.A.A., A.G.O. and I.N.A.O.; Software, A.G.O.; Validation, I.A.A.; Formal analysis, I.A.A., O.E.A., I.N.A.O. and A.G.O.; Investigation, I.A.A.; Resources, I.N.A.O. and A.G.O.; Data curation, I.A.A. and O.E.A.; Writing—original draft preparation, I.A.A.; writing—review and editing, I.A.A., A.G.O. and I.N.A.O.; Visualization, I.A.A., O.E.A. and I.N.A.O.; Supervision, A.G.O. and I.N.A.O.; Project administration, I.A.A.; Funding acquisition, I.N.A.O. and A.G.O. All authors have read and agreed to the published version of the manuscript.

Funding: The financial support of the Natural Science and Engineering Research Council (NSERC) of Canada through the Discovery Grant award (RGPIN 2017-05751) is hereby acknowledged.

Institutional Review Board Statement: Not applicable.

Informed Consent Statement: Not applicable.

Data Availability Statement: The data presented in this study are available on request from the corresponding author.

Acknowledgments: The authors thank SI Group and Washington Mills for providing the phenolic resins and SiC microfiller used in this research, respectively, at no cost.

Conflicts of Interest: The authors performed the experimental work in this study independently. The authors declare that they have no known competing financial interest or personal relationships that could have appeared to influence the work reported in this paper.

References

1. Guo, R.; Li, C.; Xian, G. Water absorption and long-term thermal and mechanical properties of carbon/glass hybrid rod for bridge cable. *Eng. Struct.* **2023**, *274*, 115176. [CrossRef]
2. Feng, B.; Wang, X.; Wu, Z.; Yang, Y.; Pan, Z. Performance of anchorage assemblies for CFRP cables under fatigue loads. *Structures* **2021**, *29*, 947–953. [CrossRef]
3. Sun, Y.; Sun, Y. Precursor infiltration and pyrolysis cycle-dependent mechanical and microwave absorption performances of continuous carbon fibers-reinforced boron-containing phenolic resins for low-density carbon-carbon composites. *Ceram. Int.* **2020**, *46*, 15167–15175. [CrossRef]
4. Nashchokin, A.V.; Malakho, A.P.; Garadzha, N.V.; Rogozin, A.D. Evolution of the physicochemical properties of carbon—Carbon composites based on phenol—Formaldehyde resins and discrete carbon fibers. *Fibre Chem.* **2016**, *47*, 465–471. [CrossRef]
5. Qanati, M.V.; Rasooli, A. Microstructural and main mechanical properties of novalac based carbon–carbon composites as the pyrolysis heating rate. *Ceram. Int.* **2021**, *47*, 26808–26821. [CrossRef]
6. De Souza, W.O.; Garcia, K.; De Avila Von Dollinger, C.F.; Pardini, L.C. Electrical behavior of carbon fiber/phenolic composite during pyrolysis. *Mater. Res.* **2015**, *18*, 1209–1216. [CrossRef]
7. Xu, Y.; Guo, L.; Zhang, H.; Zhai, H.; Ren, H. Research status, industrial application demand and prospects of phenolic resin. *RSC Adv.* **2019**, *9*, 28924–28935. [CrossRef]
8. Renda, C.G.; Bertholdo, R. Study of phenolic resin and their tendency for carbon graphitization. *J. Polym. Res.* **2018**, *25*, 241. [CrossRef]
9. Zhang, D.; Liu, X.; Bai, X.; Zhang, Y.; Wang, G.; Zhao, Y.; Li, X.; Zhu, J.; Rong, L.; Mi, C. Synthesis, characterization and properties of phthalonitrile-etherified resole resin. *E-Polymers* **2020**, *20*, 500–509. [CrossRef]
10. Bhatnagar, A. *Lightweight Ballistic Composites, Military and Law Enforcement Applications*; Woodhead Publishing Limited: Cambridge UK, 2006.
11. Duodu, E.A.; Gu, J.; Ding, W.; Shang, Z.; Tang, S. Comparison of Ballistic Impact Behavior of Carbon Fiber/Epoxy Composite and Steel Metal Structures. *Iran. J. Sci. Technol.—Trans. Mech. Eng.* **2018**, *42*, 13–22. [CrossRef]
12. Borba, N.Z.; Körbelin, J.; Fiedler, B.; dos Santos, J.F.; Amancio-Filho, S.T. Low-velocity impact response of friction riveted joints for aircraft application. *Mater. Des.* **2020**, *186*, 108369. [CrossRef]
13. Xu, F.; Zhu, S.; Liu, Y.; Ma, Z.; Li, H. Ablation behavior and mechanism of TaSi2-modified carbon fabric-reinforced phenolic composite. *J. Mater. Sci.* **2020**, *55*, 8553–8563. [CrossRef]
14. Wang, S.; Huang, H.; Tian, Y.; Huang, J. Effects of SiC content on mechanical, thermal and ablative properties of carbon/phenolic composites. *Ceram. Int.* **2020**, *46*, 16151–16156. [CrossRef]
15. Wang, S.; Huang, H.; Tian, Y. Effects of zirconium carbide content on thermal stability and ablation properties of carbon/phenolic composites. *Ceram. Int.* **2020**, *46*, 4307–4313. [CrossRef]
16. Duan, L.; Zhao, X.; Wang, Y. Oxidation and ablation behaviors of carbon fiber/phenolic resin composites modified with borosilicate glass and polycarbosilane interface. *J. Alloys Compd.* **2020**, *827*, 154277. [CrossRef]
17. Duan, L.; Luo, L.; Wang, Y. Oxidation and ablation behavior of a ceramizable resin matrix composite based on carbon fiber/phenolic resin. *Mater. Today Commun.* **2022**, *33*, 104901. [CrossRef]
18. Zhang, S.; Zhang, J.; Yang, S.; Shi, M.; Li, J.; Shen, Q. Enhanced mechanical, thermal and ablation properties of carbon fiber/BPR composites modified by mica synergistic MoSi2 at 1500 °C. *Ceram. Int.* **2023**, *49*, 21213–21221. [CrossRef]
19. Yang, T.; Dong, C.; Rong, Y.; Deng, Z.; Li, P.; Han, P.; Shi, M.; Huang, Z. Oxidation Behavior of Carbon Fibers in Ceramizable Phenolic Resin Matrix Composites at Elevated Temperatures. *Polymers* **2022**, *14*, 2785. [CrossRef] [PubMed]
20. Nagaraja, K.C.; Rajanna, S.; Prakash, G.S.; Rajeshkumar, G. Improvement of mechanical and thermal properties of hybrid composites through addition of halloysite nanoclay for light weight structural applications. *J. Ind. Text.* **2022**, *51*, 4880S–4898S. [CrossRef]
21. Lamorinière, S.; Jones, M.P.; Ho, K.; Kalinka, G.; Shaffer, M.S.P.; Bismarck, A. Carbon nanotube enhanced carbon Fibre-Poly(ether ether ketone) interfaces in model hierarchical composites. *Compos. Sci. Technol.* **2022**, *221*, 109327. [CrossRef]
22. Zhang, C.; Zhang, G.; Shi, X.P.; Wang, X. Effects of carbon nanotubes on the interlaminar shear strength and fracture toughness of carbon fiber composite laminates: A review. *J. Mater. Sci.* **2022**, *57*, 2388–2410. [CrossRef]
23. Tessema, A.; Mitchell, W.; Koohbor, B.; Ravindran, S.; Van Tooren, M.; Kidane, A. The Effect of Nano-Fillers on the In-Plane and Interlaminar Shear Properties of Carbon Fiber Reinforced Composite. *J. Dyn. Behav. Mater.* **2018**, *4*, 296–307. [CrossRef]
24. Bilisik, K.; Erdogan, G.; Karaduman, N.; Sapanci, E. Developments of Multi-nanostitched 3D Carbon/epoxy Nanocomposites: Tensile/shear and Interlaminar Properties. *Appl. Compos. Mater.* **2022**, *29*, 3–26. [CrossRef]
25. Kamae, T.; Drzal, L.T. Carbon fiber/epoxy composite property enhancement through incorporation of carbon nanotubes at the fiber-matrix interphase—Part II: Mechanical and electrical properties of carbon nanotube coated carbon fiber composites. *Compos. Part A Appl. Sci. Manuf.* **2022**, *160*, 107023. [CrossRef]
26. Lyu, H.; Jiang, N.; Li, Y.; Lee, H.P.; Zhang, D. Enhanced interfacial and mechanical properties of carbon fiber/PEEK composites by hydroxylated PEEK and carbon nanotubes. *Compos. Part A Appl. Sci. Manuf.* **2021**, *145*, 106364. [CrossRef]
27. Lyu, H.; Jiang, N.; Li, Y.; Zhang, D. Enhancing CF/PEEK interfacial adhesion by modified PEEK grafted with carbon nanotubes. *Compos. Sci. Technol.* **2021**, *210*, 108831. [CrossRef]

28. Liu, J.; Li, Y.; Xiang, D.; Zhao, C.; Wang, B.; Li, H. Enhanced Electrical Conductivity and Interlaminar Fracture Toughness of CF/EP Composites via Interleaving Conductive Thermoplastic Films. *Appl. Compos. Mater.* **2021**, *28*, 17–37. [CrossRef]
29. Kostagiannakopoulou, C.; Loutas, T.H.; Sotiriadis, G.; Kostopoulos, V. Effects of graphene geometrical characteristics to the interlaminar fracture toughness of CFRP laminates. *Eng. Fract. Mech.* **2021**, *245*, 107584. [CrossRef]
30. Liu, Y.; Li, J.; Kuang, Y.; Zhao, Y.; Wang, M.; Wang, H.; Chen, X. Interlaminar properties of carbon nanotubes modified carbon fibre fabric reinforced polyimide composites. *J. Compos. Mater.* **2023**, *57*, 1277–1287. [CrossRef]
31. Lertwassana, W.; Parnklang, T.; Mora, P.; Jubsilp, C.; Rimdusit, S. High performance aramid pulp/carbon fiber-reinforced polybenzoxazine composites as friction materials. *Compos. Part B Eng.* **2019**, *177*, 107280. [CrossRef]
32. Lai, M.; Jiang, L.; Wang, X.; Zhou, H.; Huang, Z.; Zhou, H. Effects of multi-walled carbon nanotube/graphene oxide-based sizing on interfacial and tribological properties of continuous carbon fiber/poly(ether ether ketone) composites. *Mater. Chem. Phys.* **2022**, *276*, 125344. [CrossRef]
33. Wenbin, L.; Jianfeng, H.; Jie, F.; Zhenhai, L.; Liyun, C.; Chunyan, Y. Effect of aramid pulp on improving mechanical and wet tribological properties of carbon fabric/phenolic composites. *Tribol. Int.* **2016**, *104*, 237–246. [CrossRef]
34. Rao, G.R.; Srikanth, I.; Reddy, K.L. Effect of organo-modified montmorillonite nanoclay on mechanical, thermal and ablation behavior of carbon fiber/phenolic resin composites. *Def. Technol.* **2021**, *17*, 812–820. [CrossRef]
35. Eslami, Z.; Yazdani, F.; Mirzapour, M.A. Thermal and mechanical properties of phenolic-based composites reinforced by carbon fibres and multiwall carbon nanotubes. *Compos. Part A Appl. Sci. Manuf.* **2015**, *72*, 22–31. [CrossRef]
36. Chen, Z.; Dai, X.J.; Magniez, K.; Lamb, P.R.; Rubin De Celis Leal, D.; Fox, B.L.; Wang, X. Improving the mechanical properties of epoxy using multiwalled carbon nanotubes functionalized by a novel plasma treatment. *Compos. Part A Appl. Sci. Manuf.* **2013**, *45*, 145–152. [CrossRef]
37. Feng, P.; Kong, Y.; Liu, M.; Peng, S.; Shuai, C. Dispersion strategies for low-dimensional nanomaterials and their application in biopolymer implants. *Mater. Today Nano* **2021**, *15*, 100127. [CrossRef]
38. Shi, X.; Bai, S.; Li, Y.; Yu, X.; Naito, K.; Zhang, Q. Effect of polyethylene glycol surface modified nanodiamond on properties of polylactic acid nanocomposite films. *Diam. Relat. Mater.* **2020**, *109*, 108092. [CrossRef]
39. Shin, G.J.; Kim, D.H.; Kim, J.W.; Kim, S.H.; Lee, J.H. Enhancing vertical thermal conductivity of carbon fiber reinforced polymer composites using cauliflower-shaped copper particles. *Mater. Today Commun.* **2023**, *35*, 105792. [CrossRef]
40. Cha, J.; Jin, S.; Shim, J.H.; Park, C.S.; Ryu, H.J.; Hong, S.H. Functionalization of carbon nanotubes for fabrication of CNT/epoxy nanocomposites. *Mater. Des.* **2016**, *95*, 1–8. [CrossRef]
41. Ali, A.; Rahimian Koloor, S.S.; Alshehri, A.H.; Arockiarajan, A. Carbon nanotube characteristics and enhancement effects on the mechanical features of polymer-based materials and structures—A review. *J. Mater. Res. Technol.* **2023**, *24*, 6495–6521. [CrossRef]
42. Zhang, L.; Chen, Z.; Mao, J.; Wang, S.; Zheng, Y. Quantitative evaluation of inclusion homogeneity in composites and the applications. *J. Mater. Res. Technol.* **2020**, *9*, 6790–6807. [CrossRef]
43. Marinkovic, F.S.; Popovic, D.M.; Jovanovic, J.D.; Stankovic, B.S.; Adnadjevic, B.K. Methods for quantitative determination of filler weight fraction and filler dispersion degree in polymer composites: Example of low-density polyethylene and NaA zeolite composite. *Appl. Phys. A Mater. Sci. Process.* **2019**, *125*, 611. [CrossRef]
44. Garrido-Regife, L.; Rivero-Antúnez, P.; Morales-Flórez, V. The dispersion of carbon nanotubes in composite materials studied by computer simulation of Small Angle Scattering. *Phys. B Condens. Matter* **2023**, *649*, 414450. [CrossRef]
45. Batool, M.; Haider, M.N.; Javed, T. Applications of Spectroscopic Techniques for Characterization of Polymer Nanocomposite: A Review. *J. Inorg. Organomet. Polym. Mater.* **2022**, *32*, 4478–4503. [CrossRef]
46. Zaman, A.C.; Kaya, F.; Kaya, C. A study on optimum surfactant to multiwalled carbon nanotube ratio in alcoholic stable suspensions via UV–Vis absorption spectroscopy and zeta potential analysis. *Ceram. Int.* **2020**, *46*, 29120–29129. [CrossRef]
47. Zhao, Z.; Liu, P.; Dang, H.; Nie, H.; Guo, Z.; Zhang, C.; Li, Y. Effects of loading rate and loading direction on the compressive failure behavior of a 2D triaxially braided composite. *Int. J. Impact Eng.* **2021**, *156*, 103928. [CrossRef]
48. Mohsin, M.A.A.; Iannucci, L.; Greenhalgh, E.S. On the dynamic tensile behaviour of thermoplastic composite carbon/polyamide 6.6 using split hopkinson pressure bar. *Materials* **2021**, *14*, 1653. [CrossRef]
49. Rouf, K.; Suratkar, A.; Imbert-Boyd, J.; Wood, J.; Worswick, M.; Montesano, J. Effect of strain rate on the transverse tension and compression behavior of a unidirectional non-crimp fabric carbon fiber/snap-cure epoxy composite. *Materials* **2021**, *14*, 7314. [CrossRef]
50. Lomakin, E.; Fedulov, B.; Fedorenko, A. Strain rate influence on hardening and damage characteristics of composite materials. *Acta Mech.* **2021**, *232*, 1875–1887. [CrossRef]
51. Rouf, K.; Worswick, M.J.; Montesano, J. Experimentally verified dual-scale modelling framework for predicting the strain rate-dependent nonlinear anisotropic deformation response of unidirectional non-crimp fabric composites. *Compos. Struct.* **2023**, *303*, 116384. [CrossRef]
52. Li, X.; Yan, Y.; Guo, L.; Xu, C. Effect of strain rate on the mechanical properties of carbon/epoxy composites under quasi-static and dynamic loadings. *Polym. Test.* **2016**, *52*, 254–264. [CrossRef]
53. Chihi, M.; Tarfaoui, M.; Qureshi, Y.; Bouraoui, C.; Benyahia, H. Effect of carbon nanotubes on the in-plane dynamic behavior of a carbon/epoxy composite under high strain rate compression using SHPB. *Smart Mater. Struct.* **2020**, *29*, 085012. [CrossRef]
54. Abdulganiyu, I.A.; Oguocha, I.N.A.; Odeshi, A.G. Influence of microfillers addition on the flexural properties of carbon fiber reinforced phenolic composites. *Compos. Mater.* **2021**, *55*, 3973–3988. [CrossRef]

55. Lee, Y.S.; Wetzel, E.D.; Wagner, N.J. The ballistic impact characteristics of Kevlar woven fabrics impregnated with a colloidal shear thickening fluid. *J. Mater. Sci.* **2003**, *38*, 2825–2833. [CrossRef]
56. Khan, M.M.; Iqbal, M.A. Design, development, and calibration of split Hopkinson pressure bar system for Dynamic material characterization of concrete. *Int. J. Prot. Struct.* **2023**, 1–29. [CrossRef]
57. Wang, M.; Cai, X.; Lu, Y.; Noori, A.; Chen, F.; Chen, L.; Jiang, X. Mechanical behavior and failure modes of bamboo scrimber under quasi-static and dynamic compressive loads. *Eng. Fail. Anal.* **2023**, *146*, 107006. [CrossRef]
58. Ramesh, K.T. High Rates and Impact Experiments. In *Springer Handbook of Experimental Solid Mechanics*; Sharpe, J., William, N., Eds.; Springer: New York, NY, USA, 2008; pp. 929–959.
59. Foti, G. Silicon carbide: From amorphous to crystalline material. *Appl. Surf. Sci.* **2001**, *184*, 20–26. [CrossRef]
60. Muranaka, T.; Kikuchi, Y.; Yoshizawa, T.; Shirakawa, N.; Akimitsu, J. Superconductivity in carrier-doped silicon carbide. *Sci. Technol. Adv. Mater.* **2008**, *9*, 044204. [CrossRef]
61. Anthony, J.W.; Bideaux, R.A.; Bladh, K.W.; Nichols, M.C. *Handbook of Mineralogy, Mineralogical Society of America*; Mineral Data Publishing: Chantilly, VA, USA, 1995.
62. Munasir; Triwikantoro; Zainuri, M. Darminto Synthesis of SiO_2 nanopowders containing quartz and cristobalite phases from silica sands. *Mater. Sci. Pol.* **2015**, *33*, 47–55. [CrossRef]
63. Jeon, D.; Yum, W.S.; Song, H.; Yoon, S.; Bae, Y.; Oh, J.E. Use of coal bottom ash and cao-cacl2-activated ggbfs binder in the manufacturing of artificial fine aggregates through cold-bonded pelletization. *Materials* **2020**, *13*, 5598. [CrossRef]
64. Sekhar, N.C.; Varghese, L.A. Mechanical, thermal, and rheological studies of phenolic resin modified with intercalated graphite prepared via liquid phase intercalation. *Polym. Test.* **2019**, *79*, 106010. [CrossRef]
65. Ki Park, S.; Nakhanivej, P.; Seok Yeon, J.; Ho Shin, K.; Dose, W.M.; De Volder, M.; Bae Lee, J.; Jin Kim, H.; Park, H.S. Electrochemical and structural evolution of structured V2O5 microspheres during Li-ion intercalation. *J. Energy Chem.* **2021**, *55*, 108–113. [CrossRef]

Disclaimer/Publisher's Note: The statements, opinions and data contained in all publications are solely those of the individual author(s) and contributor(s) and not of MDPI and/or the editor(s). MDPI and/or the editor(s) disclaim responsibility for any injury to people or property resulting from any ideas, methods, instructions or products referred to in the content.

Article

Research on Characterization of Nylon Composites Functional Material Filled with Al$_2$O$_3$ Particle

Jibing Chen [1,*], Bowen Liu [1], Maohui Hu [1], Qianyu Shi [1], Junsheng Chen [1], Junsheng Yang [1] and Yiping Wu [2]

1. School of Mechanical Engineering, Wuhan Polytechnic University, Wuhan 430023, China; kysdmbylbw@163.com (B.L.); hmhwhqgdx@163.com (M.H.); 15965826765@163.com (Q.S.); chenjs9610@163.com (J.C.); yangjunsheng2008@163.com (J.Y.)
2. School of Materials Science and Engineering, Huazhong University of Science and Technology, Wuhan 430074, China; ypwu@mail.hust.edu.cn
* Correspondence: jbchen@whpu.edu.cn

Abstract: This study revolves around the issues raised by the current semiconductor device metal casings (mainly composed of aluminum and its alloys), such as resource and energy consumption, complexity of the production process, and environmental pollution. To address these issues, researchers have proposed an eco-friendly and high-performance alternative material—Al$_2$O$_3$ particle-filled nylon composite functional material. This research conducted detailed characterization and analysis of the composite material through scanning electron microscopy (SEM) and differential scanning calorimetry (DSC). The results show that the Al$_2$O$_3$ particle-filled nylon composite material has a significantly superior thermal conductivity, about twice as high as that of pure nylon material. Meanwhile, the composite material has good thermal stability, maintaining its performance in high-temperature environments above 240 °C. This performance is attributed to the tight bonding interface between the Al$_2$O$_3$ particles and the nylon matrix, which not only improves the heat transfer efficiency but also significantly enhances the material's mechanical properties, with a strength of up to 53 MPa. This study is of great significance, aiming to provide a high-performance composite material that can alleviate resource consumption and environmental pollution issues, with excellent polishability, thermal conductivity, and moldability, which is expected to play a positive role in reducing resource consumption and environmental pollution problems. In terms of potential applications, Al$_2$O$_3$/PA6 composite material can be widely used in heat dissipation components for LED semiconductor lighting and other high-temperature heat dissipation components, thereby improving product performance and service life, reducing energy consumption and environmental burden, and laying a solid foundation for the development and application of future high-performance eco-friendly materials.

Keywords: PA6/Al$_2$O$_3$ composite; functional material; microstructure; mechanical property

Citation: Chen, J.; Liu, B.; Hu, M.; Shi, Q.; Chen, J.; Yang, J.; Wu, Y. Research on Characterization of Nylon Composites Functional Material Filled with Al$_2$O$_3$ Particle. *Polymers* **2023**, *15*, 2369. https://doi.org/10.3390/polym15102369

Academic Editor: S. D. Jacob Muthu

Received: 15 April 2023
Revised: 15 May 2023
Accepted: 17 May 2023
Published: 19 May 2023

Copyright: © 2023 by the authors. Licensee MDPI, Basel, Switzerland. This article is an open access article distributed under the terms and conditions of the Creative Commons Attribution (CC BY) license (https://creativecommons.org/licenses/by/4.0/).

1. Introduction

With the miniaturization, integration, high-density, and high-speed development of electronic components, heat dissipation has become a very serious problem [1], especially as various light-emitting diodes (LEDs) are widely used in various places for urban lighting. High-power and high-brightness LEDs generate a tremendous amount of heat, making it challenging to maintain the optimal operating temperature range of 25–30 °C for LEDs. Traditional inorganic thermally conductive materials such as metals and ceramics can no longer meet the needs of further development of electronic products, and new high-efficiency thermally conductive materials must be used to dissipate heat promptly [2]. Polymer-based composite materials refer to the introduction of one or more inorganic or organic materials with specific properties into the polymer matrix through various processing methods (including chemical and physical means) so that the polymer also has some special properties, such as thermal conductivity, shielding of electromagnetic waves, dielectric properties, electrical conductivity, and excellent damping performance [3]. Thermally conductive polymer composites have excellent thermal conductivity and unparalleled molding processability, electrical insulation, and corrosion resistance of traditional inorganic thermally conductive materials [4], becoming a research hotspot in the field of polymer materials.

The thermal conductivity of metals mainly relies on freely moving electrons, and the heat transfer mechanism is the formation of a heat conduction band using unconstrained electrons that can move freely; the thermal conductivity of inorganic non-metallic materials can only be conducted through the microscopic structure of mutual contact between material molecules by transferring phonons, using phonons as heat transfer carriers to transfer heat. Phonons are mechanical waves generated by lattice vibrations in crystals (mainly vibrations of atoms, molecules, and groups) [5,6], which are successively transferred between material molecules, so their thermal conductivity efficiency is not as good as that of free electrons. In contrast, polymers themselves are composed of molecular chains with very large molecular weights, and it is difficult to form a free electron energy band inside the material. Therefore, polymers mainly transfer heat through mutual contact between microscopic structures. However, unlike the phonon thermal conductivity of inorganic non-metallic materials, polymers themselves are composed of mixtures of homologous substances with different molecular weights, with larger van der Waals forces between molecular chains and longer chains that are prone to entanglement at the ends, resulting in more defects between the molecules. Therefore, polymer materials are particularly prone to phonon scattering, leading to low heat transfer efficiency and poor thermal conductivity of polymer materials [7]. Currently, there are two main approaches to preparing thermally conductive composite materials [8]: intrinsically thermally conductive polymer composites with inherent heat transfer capabilities, and filler-based thermally conductive polymer composites using externally added fillers. Intrinsic thermally conductive polymer materials have a large number of conjugated structures in their molecular chains and a relatively large proportion of crystalline regions, and these special structures give the polymer itself good heat transfer capabilities [9]. When phonons propagate in these polymers, there is less phonon scattering and hindrance, making intrinsic thermally conductive polymer composites excellent thermal conductors. However, the processing technology of intrinsic thermally conductive polymers is very complex, from the selection of initiators and the initiation of active polymer monomers, to the influence of reaction temperature and reaction time. These factors make the preparation of intrinsic thermally conductive polymers extremely difficult and the industrial production and processing difficult [10]. Therefore, based on the modern Fourier solid heat conduction theory, inorganic powders with strong heat transfer capabilities are filled into polymers to prepare high thermally conductive base polymer materials that meet the needs of the actual operating environment.

At present, a large number of studies have been conducted both domestically and abroad, aiming to develop composite functional materials with high thermal conductivity. Joao Paulo Berenguer, Arielle Berman, and other scholars [11] studied the effects of polyethylene fillers on the thermal conductivity, mechanical properties, and processability of all-polymer high thermal conductivity composite materials. Through thermal conductivity tests, tensile tests, bending tests, and impact tests, the results showed that optimizing the mass fraction and dispersion degree of polyethylene fillers can achieve a comprehensive optimization of the thermal conductivity, mechanical properties, and processability of all-polymer high thermal conductivity composite materials. Guorui Zhang, Sen Xue, and others [12] developed a high-efficiency thermal interface material with anisotropic orientation and high vertical thermal conductivity, by adjusting the distribution and orientation of graphene fillers, improving the thermal management performance of microelectronic devices, extending device life, and improving energy efficiency. You Y-L, Li D-X, and others [13] prepared GF/PA6 composite materials containing different solid lubricants in the laboratory, and used a ball-disc friction tester to evaluate the friction and wear performance of different composite materials, revealing the mechanism of various solid lubricants in reducing friction and wear. One paper [14] mainly studied the thermal conductivity of polymer composite materials based on phenolic resin and boron nitride (BN), by adding boron nitride fillers to phenolic resin, improving the thermal conductivity of composite materials, and evaluating the thermal conductivity, thermal stability, and microstructure of composite materials through thermal conductivity tests, thermogravimetric analysis, and scanning electron microscopy. Another [15] analyzed the microstructure, thermal stability, and mechanical properties of Al_2O_3 and SiC-reinforced PA6 hybrid composite materials, and the experimental results showed that Al_2O_3 and SiC particles were uniformly distributed in the PA6 matrix, effectively improving the mechanical properties of the materials. Jiaqi Zhang, Xianzhao Jia, and others [16] mainly studied the effects of additive fluoroelastomer (FVMQ) and Al_2O_3 particle co-filling on its mechanical properties, thermal properties, and friction properties at high temperatures, verifying that by optimizing the mass fraction and dispersion degree of Al_2O_3 particle fillers, a comprehensive optimization of the mechanical properties, thermal properties, and friction properties of fluoroelastomer composite materials under high-temperature conditions can be achieved. Qiu Hong Mu, Dan Peng, and other scholars [17] studied the effect of filling Al_2O_3 particles on the thermal conductivity of silicone rubber, and evaluated the thermal conductivity, thermal stability, and microstructure of composite materials through thermal conductivity tests, thermogravimetric analysis, and scanning electron microscopy. One work [18] explored the effects of filling Al_2O_3 compounds on the thermal conductivity and rheological properties of epoxy resin and liquid crystal epoxy resin composite materials, by adding Al_2O_3 compound fillers to epoxy resin and liquid crystal epoxy resin, improving the thermal conductivity and rheological properties of composite materials. Konopka K, Krasnowski M, and others [19] used pulsed plasma sintering (PPS) method to prepare Al_2O_3 samples and $NiAl-Al_2O_3$ composite materials, and analyzed the microstructure, phase composition, thermal stability, and mechanical properties of the two materials by X-ray diffraction (XRD), thermogravimetric analysis (TGA), and differential scanning calorimetry (DSC), and evaluated the significant influence of pulsed plasma sintering parameters on the microstructure and mechanical properties

According to the literature cited above, a series of research work has been carried out on thermal conductive materials for semiconductor devices. In order to ensure that these thermal conductive materials have excellent comprehensive performance, researchers usually choose to fill polymer materials with high thermal conductivity inorganic fillers or metal fillers. The thermal conductive materials prepared in this way have the advantages of low cost and easy processing. Through appropriate physical and chemical treatment methods or adjusting the experimental formula ratio, such thermal conductive materials can meet the specific application scenarios with thermal conductivity requirements. PA6 has good mechanical properties, thermal stability, and chemical stability, and can adapt to various complex working environments. More importantly, PA6 has high adjustability, and its performance can be improved by adding different types and proportions of fillers, reinforcements, and auxiliaries. Therefore, this study chose to add inorganic fillers (Al_2O_3) to nylon (PA6) and add 5% diethylhexyl phthalate (DEHP) in a certain proportion to prepare materials with specific functions. In order to study the mechanical properties of these functional materials, various instruments such as scanning electron microscopy (SEM) (JSM-7500F, JEOL Ltd., Tokyo, Japan), differential scanning calorimeter (DSC) (1/700 type, Mettler-Toledo, Zurich, Switzerland), and thermal conductivity meter (DTC-300, TA Instruments, New Castle, DE, USA) were used for testing and analysis. The results show that this high thermal conductivity functional material has great potential to replace metal shell materials and can be used to manufacture semiconductor devices with a metal appearance. This material performs excellently in polishing performance, thermal conductivity, and molding performance, providing promising possibilities for replacing metal materials. For example, in LED semiconductor lighting, it can be used as an effective heat dissipation material.

2. Materials and Methods

2.1. Raw Materials

The thermally conductive base material nylon (PA6) was purchased from Zhuhai SMIKA Polymer Company (Zhuhai, China); the thermally conductive filler Al_2O_3 (purity $\geq$ 99%) was obtained from National Medicine Group Chemical Reagent Co., Ltd. (Shanghai, China); the coupling agent diethylhexyl phthalate (DEHP) was provided by Foshan New Material Technology Co., Ltd. (Foshan, China).

2.2. Preparation of Materials

Initially, the nylon plastic granules underwent essential pretreatment. The nylon granules were dried in an oven at a temperature of 120 °C for 24 h to eliminate moisture, ensuring the stability of material performance in subsequent experiments.

In this study, we conducted three sets of experiments. Firstly, to achieve a good thermal conductivity, thermal stability, and mechanical properties, we combined the research from ref. [20,21] and prior theoretical analysis. Nylon PA6, thermal conductive filler Al_2O_3, and DEHP were mixed at a mass ratio of 80:18:2. A high-speed stirrer was used to ensure that all components were fully mixed and formed a homogeneous mixture. Subsequently, the obtained mixture was extruded using an extruder to give the material a certain shape. The extruded material was then dried at 80 °C for 2 h to remove any residual moisture, ensuring the smooth progress of the subsequent injection molding process.

Secondly, with the aforementioned experimental procedure kept consistent, we conducted grouped experiments with filler particle size, filler shape, and filler ratio as variables to evaluate the thermal performance of the material. The particle sizes selected were 5 µm, 20 µm, and 100 µm; the filler shapes were spherical and flaky; the filler ratios were 10%, 20%, 30%, 40%, and 50%.

Lastly, to analyze the relationship between the filler ratio and mechanical properties in more detail, we set up nine groups of gradient experiments for the filler ratio, namely 0%, 5%, 10%, 15%, 20%, 25%, 30%, 35%, and 40%, while maintaining the aforementioned experimental procedure.

2.3. Characterization of Thermal Conductivity

After the experiment was completed, we immersed the samples used for testing thermal performance in liquid nitrogen, allowing them to rapidly solidify and become brittle at low temperatures, thus facilitating subsequent cross-sectional observation and analysis. Then, the cross-section of the sample was placed on conductive adhesive and a sputter coating treatment performed to provide good conductivity and adhesion for observation and analysis under a scanning electron microscope (SEM). The SEM was used to observe the distribution characteristics and morphology of the thermally conductive filler in the matrix material. Finally, the prepared heat-resistant plastic samples were sent to a differential scanning calorimeter (DSC) for heat resistance performance analysis. We sought to understand the thermal stability and thermal degradation behavior of the composite material at different temperatures, thereby evaluating its application potential in high-temperature environments. Furthermore, we employed a thermal conductivity meter to analyze the samples. By applying heat to the sample and measuring its temperature change over a certain period, we calculated the sample's thermal conductivity, which allowed us to predict the material's thermal management performance in practical applications.

2.4. Mechanical Analysis

In this study, tensile tests were conducted to evaluate the mechanical properties of the composite material, using an electronic universal testing machine (WDT-10, Shenzhen Kaiqiangli Experimental Instrument Co., Ltd., Shenzhen, China.) with a testing temperature of 25 °C. To ensure the accuracy and comparability of the experiments, the preparation and testing process of the tensile samples followed the relevant provisions of ASTM D638. Meanwhile, the size of the samples referred to the ISO 527-2 Type 1A standard requirements to ensure that the shape and dimensions of the samples met internationally accepted requirements. In the actual testing process, the samples were stretched at a transverse speed of 25 mm/min to simulate the loading speed and conditions that may be encountered in real-life applications. Finally, the American TA Instruments (New Castle, DE, USA) Q800 analyzer was used for single cantilever mode testing, gaining a deeper understanding of the mechanical properties of the composite material and its potential in various application scenarios.

2.5. Scanning Electron Microscopy (SEM)

A scanning electron microscope (SEM) was used to study the microstructure of PA6, Al_2O_3, and the composite material. Observing the microstructure and interactions within PA6, Al_2O_3, and their composite material helps to reveal the microstructural characteristics of the samples.

3. Results and Discussion

3.1. Microstructure Analysis of Thermal Polymer Materials

In order to measure the thermal conductivity, the test samples were first treated with liquid nitrogen to increase the brittleness of the material. Subsequently, the cross-section of the samples was placed on a conductive adhesive and subjected to sputter coating, enabling observation of the distribution characteristics and morphology of the thermally conductive fillers within the matrix material under a scanning electron microscope (SEM). Following this step, we utilized a differential scanning calorimeter (DSC) to analyze the thermal stability of the heat-resistant plastic samples to understand their performance under high-temperature conditions. Prior to the formal preparation of thermally conductive polymer materials, a detailed examination of the microstructure and energy spectrum distribution of the fillers and PA6 was conducted using a scanning electron microscope, based on the composition ratio of the thermally conductive fillers. The results are shown in Figures 1 and 2.

Figure 1. PA6 material microstructure image.

After successfully preparing the thermally conductive composite material, its microstructure was observed using a scanning electron microscope (SEM), as shown in Figure 3. Comparing Figures 1 and 3, it can be found that the SEM image of the original PA6 displays a relatively uniform and smooth surface, while the SEM image of the PA6 composite material with added Al_2O_3 filler exhibits a clear distribution of filler particles and an interfacial region is formed between the Al_2O_3 filler and the PA6 matrix. Furthermore, the thermally conductive filler (Al_2O_3) appears as flake-like or spherical shapes and is relatively uniformly distributed within the nylon matrix material, which is beneficial for forming continuous thermal conduction paths or networks within the composite material. After the addition of the thermally conductive filler, the original microstructure of the matrix material changes, resulting in the thermally conductive composite material exhibiting excellent thermal conductivity performance.

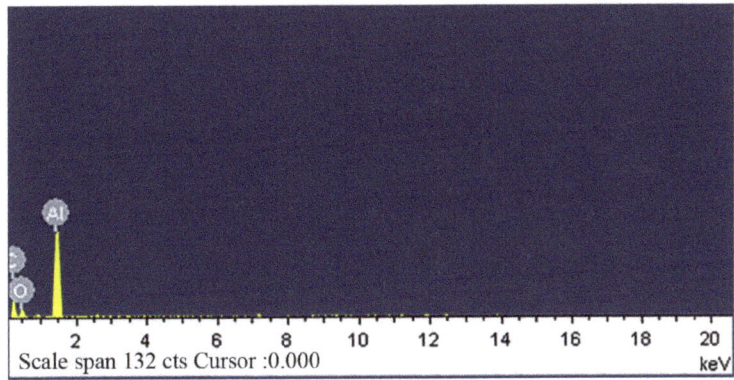

Figure 2. Al$_2$O$_3$ energy spectrum analysis image.

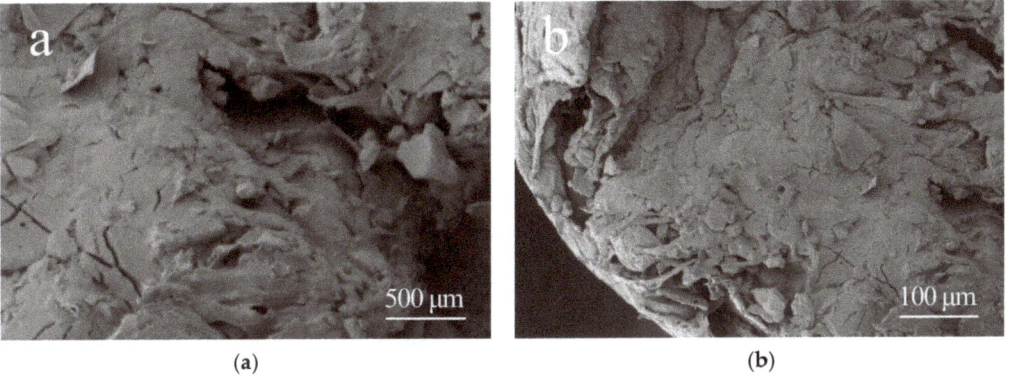

Figure 3. Figure 3. SEM photographs of nylon composites: (**a**) 500 μm; (**b**) 100 μm.

3.2. DSC Testing

In this study, the thermal properties of the thermally conductive polymer samples were analyzed using a differential scanning calorimeter (DSC). During the experiment, the samples were placed in a nitrogen atmosphere and heated at a rate of 20 °C per minute, with the temperature increasing from room temperature to 300 °C. After reaching 300 °C, the temperature was maintained at 10 °C for 10 min, followed by cooling. The obtained DSC curve of the samples is shown in Figure 4. Based on the results in Figure 4, the entire DSC curve can be divided into four stages [22]. In the first stage, from the start of heating to approximately 160 °C, the DSC curve gradually rises and reaches the peak of the first endothermic peak; at this temperature, the polymer transitions from a glassy state to a highly elastic rubber-like or viscous fluid state, and the motion of the molecular chains becomes more flexible. The subsequent second stage extends from the end of the 160 °C endothermic peak to a temperature of 175 °C. The third stage has a temperature range of 175 °C to 240 °C, where the curve starts to rise, ultimately forming the second endothermic peak; at this temperature, the crystalline regions of PA6 begin to melt, and the movement of the molecular chains further increases. The final fourth stage covers a temperature range of 230 °C to 300 °C, during which the DSC curve begins to decline and exhibits a faster descent rate; in this process, heat is released, and the polymer material's chain segments begin to rearrange and form crystalline regions.

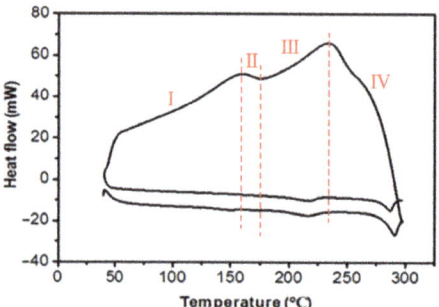

Figure 4. DSC curves of thermally conductive polymers. (I) Glass transition phase; (II) Thermal stabilization phase; (III) Melting phase; (IV) Crystallization phase.

3.3. Thermal Conductivity Testing

In this project, we conducted experiments on the thermal conductivity of polymer materials prepared based on Al_2O_3 with different diameters and morphologies. The diameters of the fillers used were 5 µm, 20 µm, and 100 µm, and the morphologies included spherical and flake-like shapes. Figure 5 shows the corresponding experimental data curves. Combining the experimental data with the literature [23], several conclusions can be drawn: (1) Under different diameters or different morphologies of the filler, the thermal conductivity of the polymer materials exhibits an increasing trend, indicating that the thermal conductivity of nylon composites increases with the increase in the content of Al_2O_3 filler. This phenomenon can be attributed to the increase in filler content, which helps form thermal conduction paths and networks in the composite material, thereby improving thermal conductivity; (2) By observing the data curves of the three groups with filler diameters of 5 µm, 20 µm, and 100 µm, it can be concluded that the thermal conductivity is the highest when the filler diameter is 20 µm and the lowest when the filler diameter is 100 µm. This indicates that within a specific particle size range, the thermal conductivity of the composite material increases with the increase in particle size; however, beyond this range, the thermal conductivity gradually decreases. This phenomenon may be due to the larger particle size filler reducing the thermal conductivity of the composite material, leading to a decrease in overall thermal conductivity; (3) Comparing the data of spherical and flake-like fillers with a diameter of 5 µm, it can be found that the flake-like fillers are more effective in improving the thermal conductivity of the composite material than spherical fillers.

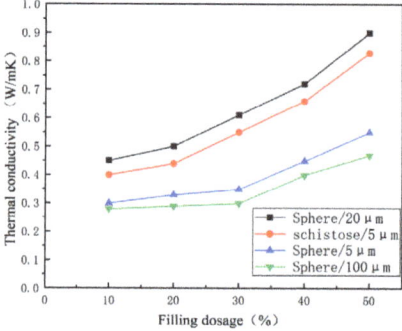

Figure 5. The thermal conductivity of spherical and flaky Al_2O_3 filled with nylon composites with 5, 20, and 100 µm particles.

3.4. Mechanical Properties Testing

Figure 6 shows the change curve of the tensile strength of composite materials as the content of Al_2O_3 filler increases. According to the experimental data curve, the tensile strength of nylon is 46 MPa. When the alumina content reaches 14%, the tensile strength of the composite material reaches its maximum value of 55.1 MPa, which is about 19% higher than that of pure nylon material. However, when the alumina filler content exceeds 14%, the tensile strength of the composite material significantly decreases, showing an inverse relationship with the proportion of filler. When the alumina content reaches 30%, the tensile strength is 35.5 MPa. This observation indicates that the tensile strength of the composite material increases with the increase in alumina content within a specific range (alumina content below 10%). When the content is below this range, its tensile strength is even lower than that of single nylon material.

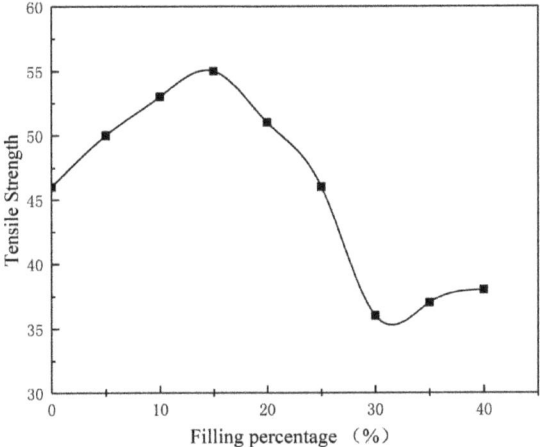

Figure 6. Tensile strength curves of alumina/nylon composites. ■ denotes the tensile strength at various filler percentages.

With the addition of alumina filler, the compressive strength of nylon composite materials has been improved accordingly. An electronic universal testing machine was used to measure the compressive strength of composite materials, and the experimental results are shown in Figure 7. When the alumina content reaches 15%, the compressive strength of the composite material reaches its highest value of 94 MPa. Compared with pure PA6, the compressive strength of this composite material has increased by 47.6%. Within a certain range (alumina content $\leq$ 15%), the compressive strength of nylon composite materials increases with the increase in alumina content. This trend can be attributed to the dispersion of alumina particles in the polymer matrix and the interaction between alumina particles and the polymer matrix [24]. However, when the alumina content exceeds this threshold, the excessive filler content leads to particle aggregation in the polymer matrix, which affects the overall performance of the composite material, resulting in a decreasing trend in compressive strength.

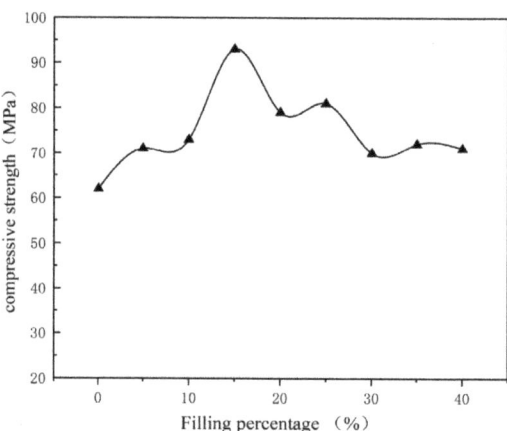

Figure 7. Compressive strength curve of Al$_2$O$_3$/nylon composites. ▲ denotes the compressive strength at various filler percentages.

4. Conclusions

In this paper, researchers successfully prepared a composite material with high thermal conductivity using high thermal conductive functional materials, namely polyamide 6 (PA6) and alumina (Al$_2$O$_3$). Detailed experimental studies were conducted on the thermal conductivity, heat resistance, and performance at different filler ratios of this composite material. The experimental results show that: (1) using spherical fillers with a diameter of 20 μm can achieve a larger thermal conductivity. When the filler size is reduced to 5 μm, the thermal conductivity reaches the highest value of 0.9 W·(mK)$^{-1}$. In addition, this composite material has good heat resistance performance and can maintain stability at the highest temperature of 240 °C. (2) The composite material has significant mechanical properties. When the Al$_2$O$_3$ content is 10%, the tensile strength of the thermal conductive functional material reaches its highest value. However, when the additive amount is increased to 14%, the compressive strength of the composite material reaches the highest value of 55.1 MPa, indicating that to a certain extent, increasing the Al$_2$O$_3$ content can improve the mechanical properties of the composite material. In summary, this high thermal conductive functional material prepared from PA6 and Al$_2$O$_3$ not only has excellent thermal conductivity and heat resistance performance but also has good mechanical properties. This gives this composite material a wide range of application potential in high-temperature application fields, such as aerospace, automotive manufacturing, and electronic devices. This research provides beneficial practical experience for engineers and scientists in related fields, helping to develop higher performance composite materials in actual applications.

Author Contributions: Conceptualization, J.C. (Jibing Chen) and B.L.; methodology, J.C. (Jibing Chen); software, J.C. (Jibing Chen) and B.L.; validation, J.C. (Junsheng Chen), B.L. and M.H.; formal analysis, J.Y. and Y.W.; investigation, J.C. (Junsheng Chen) and B.L.; resources, J.C. (Junsheng Chen); data curation, M.H. and Q.S.; writing—original draft preparation, B.L., J.C. (Junsheng Chen) and M.H.; writing—review and editing, Q.S. and Y.W.; supervision, J.C. (Jibing Chen), Y.W. and J.Y.; project administration, J.C. (Jibing Chen) and Y.W.; funding acquisition, J.C. (Jibing Chen) and J.Y. All authors have read and agreed to the published version of the manuscript.

Funding: This research was supported by the Open Fund of Hubei Longzhong Laboratory (2022ZZ-17) and the Science and Technology Project of Science and Technology Department of Hubei Province (No. 2022EHB020).

Institutional Review Board Statement: Not applicable.

Informed Consent Statement: Not applicable.

Data Availability Statement: Not applicable.

Acknowledgments: The authors acknowledge the technical support of Hubei Longzhong Laboratory, and the Analytical and Testing Center of HUST for their analytical work.

Conflicts of Interest: The authors declare no conflict of interest.

References

1. Zhang, X.; Gai, P.; Zhang, B.; Tang, Y. Thermal conductivity of rubber composite materials with a hybrid AlN/carbon fiber filler. *Chin. Sci. Bull.* **2018**, *63*, 2403–2410. [CrossRef]
2. Houshyar, S.; Padhye, R.; Troynikov, O.; Nayak, R.; Ranjan, S. Evaluation and improvement of thermo-physiological comfort properties of firefighters' protective clothing containing super absorbent materials. *J. Text. Inst.* **2015**, *106*, 1394–1402. [CrossRef]
3. Abbasov, H. The effective thermal conductivity of polymer composites filled with high conductive particles and the shell structure. *Polym. Compos.* **2022**, *43*, 2593–2601. [CrossRef]
4. Shao, Z.; He, X.; Niu, Z.; Huang, T.; Cheng, X.; Zhang, Y. Ambient pressure dried shape-controllable sodium silicate based composite silica aerogel monoliths. *Mater. Chem. Phys.* **2015**, *162*, 346–353. [CrossRef]
5. Chen, H.; Sui, X.; Zhou, C.; Wang, C.H.; Yin, C.X.; Liu, F.T. Preparation and characterization of mullite fiber-reinforced Al_2O_3-SiO_2 aerogel composites. *Key Eng. Mater.* **2016**, *697*, 360–363. [CrossRef]
6. Zhao, Y.; Zeng, X.; Ren, L.; Xia, X.; Zeng, X.; Zhou, J. Heat conduction of electrons and phonons in thermal interface materials. *Mater. Chem. Front.* **2021**, *5*, 5617–5638. [CrossRef]
7. Mehra, N.; Mu, L.; Zhu, J. Developing heat conduction pathways through short polymer chains in a hydrogen bonded polymer system. *Compos. Sci. Technol.* **2017**, *148*, 97–105. [CrossRef]
8. Liu, C.; Rao, Z.; Zhao, J.; Huo, Y.; Li, Y. Review on nanoencapsulated phase change materials: Preparation, characterization and heat transfer enhancement. *Nano Energy* **2015**, *13*, 814–826. [CrossRef]
9. Jasmee, S.; Omar, G.; Othaman, S.S.C.; Masripan, N.A.; Hamid, H.A. Interface thermal resistance and thermal conductivity of polymer composites at different types, shapes, and sizes of fillers: A review. *Polym. Compos.* **2021**, *42*, 2629–2652. [CrossRef]
10. Bose, P.; Amirtham, V.A. A review on thermal conductivity enhancement of paraffinwax as latent heat energy storage material. *Renew. Sustain. Energy Rev.* **2016**, *65*, 81–100. [CrossRef]
11. Berenguer, J.P.; Berman, A.; Quill, T.; Zhou, T.; Kalaitzidou, K.; Cola, B.; Bougher, T. Incorporation of polyethylene fillers in all-polymer high-thermal-conductivity composites. *Polym. Bull.* **2021**, *78*, 3835–3850. [CrossRef]
12. Zhang, G.; Xue, S.; Chen, F.; Fu, Q. An efficient thermal interface material with anisotropy orientation and high through-plane thermal conductivity. *Compos. Sci. Technol.* **2022**, *231*, 109784. [CrossRef]
13. You, Y.-L.; Li, D.-X.; Deng, X.; Li, W.J.; Xie, Y. Effect of solid lubricants on tribological behavior of glass fiber reinforced polyamide 6. *Polym. Compos.* **2013**, *34*, 1783–1793. [CrossRef]
14. Samoilov, V.M.; Danilov, E.A.; Kaplan, I.M.; Lebedeva, M.V.; Yashtulov, N.A. Thermal Conductivity of Polymer Composite Material Based on Phenol-Formaldehyde Resin and Boron Nitride. *Russ. Phys. J.* **2022**, *65*, 80–90. [CrossRef]
15. Sathees Kumar, S.; Kanagaraj, G. Investigation of Characterization and Mechanical Performances of Al_2O_3 and SiC Reinforced PA6 Hybrid Composites. *J. Inorg. Organomet. Polym. Mater.* **2016**, *26*, 788–798. [CrossRef]
16. Zhang, J.; Jia, X.; He, Q. Mechanical, thermal, and friction properties of addition-type fluororubber co-filled with Al_2O_3 particles at high temperature. *Polym. Test.* **2021**, *96*, 107131. [CrossRef]
17. Mu, Q.H.; Peng, D.; Wang, F.; Li, J.H.; Zhang, S. Thermal Conductivity of Silicone Rubber Filled with Al_2O_3. *Mater. Sci. Forum* **2020**, *987*, 59–63. [CrossRef]
18. Tian, F.; Cao, J.; Ma, W. Enhanced thermal conductivity and rheological performance of epoxy and liquid crystal epoxy composites with filled Al_2O_3 compound. *Polym. Test.* **2023**, *120*, 107940. [CrossRef]
19. Konopka, K.; Krasnowski, M.; Zygmuntowicz, J.; Cymerman, K.; Wachowski, M.; Piotrkiewicz, P. Characterization of Al_2O_3 Samples and NiAl–Al_2O_3 Composite Consolidated by Pulse Plasma Sintering. *Materials* **2021**, *14*, 3398. [CrossRef]
20. Satheeskumar, S.; Kanagaraj, G. Experimental investigation on tribological behaviours of PA6, PA6-reinforced Al_2O_3 and PA6-reinforced graphite polymer composites. *Bull. Mater. Sci.* **2016**, *39*, 1467–1481. [CrossRef]
21. Wieme, T.; Augustyns, B.; Duan, L.; Cardon, L. Increased through-plane thermal conductivity of injection moulded thermoplastic composites by manipulation of filler orientation. *Plast. Rubber Compos.* **2022**, *51*, 110–117. [CrossRef]
22. Abd-El-Galiel, D.; El-Matarawy, A.; El-Dien, E.M. Determination of Al_2O_3 Powder Thermal Conductivity Using DSC at NIS-Egypt. *J. Adv. Res. Fluid Mech. Therm. Sci.* **2022**, *97*, 149–156. [CrossRef]
23. Li, H.; Zheng, W. Enhanced thermal conductivity of epoxy/alumina composite through multiscale-disperse packing. *J. Compos. Mater.* **2021**, *55*, 17–25. [CrossRef]
24. Shanmugan, S.; Nurjassriatul, J.A.; Mutharasu, D. Chemical Vapor Deposited Al_2O_3 Thin Film as Thermal Interface Material for High Power LED Application. *J. Optoelectron. Biomed. Mater.* **2016**, *8*, 137–146.

Disclaimer/Publisher's Note: The statements, opinions and data contained in all publications are solely those of the individual author(s) and contributor(s) and not of MDPI and/or the editor(s). MDPI and/or the editor(s) disclaim responsibility for any injury to people or property resulting from any ideas, methods, instructions or products referred to in the content.

Article

Morphological, Mechanical, and Antimicrobial Properties of PBAT/Poly(methyl methacrylate-*co*-maleic anhydride)–SiO$_2$ Composite Films for Food Packaging Applications

Raja Venkatesan [1,†], Krishnapandi Alagumalai [1,†], Chaitany Jayprakash Raorane [1], Vinit Raj [1], Divya Shastri [2,*] and Seong-Cheol Kim [1,*]

[1] School of Chemical Engineering, Yeungnam University, Gyeongsan 38541, Republic of Korea
[2] School of Pharmacy, Yeungnam University, Gyeongsan 38541, Republic of Korea
* Correspondence: divyashastri8@gmail.com (D.S.); sckim07@ynu.ac.kr (S.-C.K.); Tel.: +82-53-810-2787 (S.-C.K.)
† These authors contributed equally to this work.

Abstract: A poly(methyl methacrylate-*co*-maleic anhydride) P(MMA-*co*-MA) copolymer was synthesized via radical polymerization. The synthesized P(MMA-*co*-MA) copolymer was identified by ^{1}H- and ^{13}C-nuclear magnetic resonance spectroscopy (^{1}H-NMR), (^{13}C-NMR), Fourier-transform infrared (FTIR) spectroscopy, X-ray diffraction (XRD), scanning electron microscopy (SEM), and transmission electron microscopy (TEM). The poly(butylene adipate-*co*-terephthalate) (PBAT)/P(MMA-*co*-MA)–SiO$_2$ composites were developed using a solution-casting method. The PBAT to P(MMA-*co*-MA) weight ratio was kept at 70:30, while the weight percentage of SiO$_2$ nanoparticles (NPs) was varied from 0.0 to 5.0 wt.%. SiO$_2$ was used for PBAT/P(MMA-*co*-MA) to solve the compatibility between PBAT and the P(MMA-*co*-MA) matrix. The PBAT/P(MMA-*co*-MA)–SiO$_2$ composites were characterized by studied FTIR spectroscopy, XRD, SEM, and TEM. A comparison of the composite film PBAT/P(MMA-*co*-MA)–SiO$_2$ (PBMS-3) with the virgin PBAT and P(MMA-*co*-MA) film revealed its good tensile strength (19.81 MPa). The WVTR and OTR for the PBAT/P(MMA-*co*-MA)–SiO$_2$ composites were much smaller than for PBAT/P(MMA-*co*-MA). The PBAT/P(MMA-*co*-MA)–SiO$_2$ WVTR and OTR values of the composites were 318.9 ± 2.0 (cc m^{-2} per 24 h) and 26.3 ± 2.5 (g m^{-2} per 24 h). The hydrophobicity of the PBAT/P(MMA-*co*-MA) blend and PBAT/P(MMA-*co*-MA)–SiO$_2$ composites was strengthened by the introduction of SiO$_2$, as measured by the water contact angle. The PBAT/P(MMA-*co*-MA)–SiO$_2$ composite films showed excellent antimicrobial activity against the food-pathogenic bacteria *E. coli* and *S. aureus* from the area of inhibition. Overall, the improved packaging characteristics, such as flexibility, tensile strength, low O$_2$ and H$_2$O transmission rate, and good antimicrobial activities, give the PBAT/P(MMA-*co*-MA)–SiO$_2$ composite film potential for use in food packaging applications.

Keywords: PBAT; P(MMA-*co*-MA) copolymer; SiO$_2$ nanoparticles; PBAT/P(MMA-*co*-MA)–SiO$_2$ composites; mechanical strength; antimicrobial activity; food packaging

Citation: Venkatesan, R.; Alagumalai, K.; Raorane, C.J.; Raj, V.; Shastri, D.; Kim, S.-C. Morphological, Mechanical, and Antimicrobial Properties of PBAT/Poly(methyl methacrylate-*co*-maleic anhydride)–SiO$_2$ Composite Films for Food Packaging Applications. *Polymers* **2023**, *15*, 101. https://doi.org/10.3390/polym15010101

Academic Editor: S. D. Jacob Muthu

Received: 17 November 2022
Revised: 17 December 2022
Accepted: 21 December 2022
Published: 26 December 2022

Copyright: © 2022 by the authors. Licensee MDPI, Basel, Switzerland. This article is an open access article distributed under the terms and conditions of the Creative Commons Attribution (CC BY) license (https://creativecommons.org/licenses/by/4.0/).

1. Introduction

The advances in food packaging materials in recent decades have attracted considerable attention because of the increasing need for effectively packaged foods with an extended shelf life [1]. The active film can offer extra safety features in addition to the usual barrier characteristics [2,3]. The antioxidant and antimicrobial activities are significant because they help alleviate concerns with food quality, such as degradation, rancidity, and coloration. An active food packaging material should fulfill a list of needs, including having sufficient physical and mechanical properties, a high barrier, a high level of barrier performance, safe for human health, and no harmful impact on the environment [4]. Furthermore, a straightforward, reproducible, and economical preparation method is essential for large-scale applications [5,6]. New production techniques that enhance the safety of food packaging are available. The manufacture of food packaging must ensure that the packaging volume and weight are kept to a minimum to maintain the required stability, hygienic conditions, and consumer acceptance. However, there are significant obstacles to using biodegradable food packaging in markets. Technologies that enhance the oxygen and water barrier of biodegradable polymer systems are ideal for biodegradable packaging in daily life. In recent times, notable potential improvements have been adopted, showing promise in fabricating high oxygen/water vapor-permeable biodegradable materials for food packaging.

Radical polymerization is the primary method for synthesizing poly(methyl methacrylate-co-maleic anhydride) P(MMA-co-MA) [7]. Pre-polymerization and cast polymerization are the two steps that are always involved in polymerization [8]. Furthermore, P(MMA-co-MA) copolymer can be used in ring-opening reactions to form cube-shaped materials that can be used in various applications, such as electrolytes for batteries, surfactants and emulsified solutions, and antiwear additives [9,10]. It is often used to fabricate composite materials that transfer heat easily because of its excellent properties [11–13]. P(MMA-co-MA) is the focus of the current research, which shows a relationship with particle size [14], flame retardants [15], inorganic nanoparticles [16], bio-additives [17], chemical modification [18], and polymerization conditions [19]. On the other hand, the brittleness and poor mechanical characteristics of P(MMA-co-MA) constitute a significant impediment to its broad application. Some research has been conducted to develop P(MMA-co-MA) composites with reinforcement materials, such as polymers, chitin nanofibers, and cellulose, to increase the mechanical and thermal properties and solve the issues [11]. Nevertheless, there are still a few disadvantages. For example, it cannot substantially enhance the mechanical properties of P(MMA-co-MA) materials, and using nanofillers can increase the cost of P(MMA-co-MA) composites, which is another impediment to scaling up production. Impurities cause opacity and the use is confined because of the hydrolysis reaction of the maleic anhydride groups [20,21]. Maleic anhydride (MA) and methyl methacrylate (MMA) were copolymerized. Other components may alter P(MMA-co-MA) to circumvent these limitations [12,22].

As a transparent thermoplastic, lightweight, or shatter-resistant replacement for glassware, poly(butylene adipate-co-terephthalate) (PBAT) was blended with P(MMA-co-MA). It is non-toxic and contains a strong hydrogen bonding character. It can also be used as a casting polymer in inks and coatings. Owing to its plasticizing property [9], PBAT increases the flexibility and barrier characteristics of the produced composite film, which are essential for packaging applications [23,24]. The most notable biocompatibility polymers in the thermoplastic materials group, PBAT, are synthesized via reactions among the monomers adipic acid, terephthalic acid, and 1,4-butanediol [25]. This material is attractive because of its excellent physical features, high elongation at break, and amorphous structure. In addition, it can be obtained using eco-friendly and renewable technologies [26,27]. The capability to produce scaffolding structures enables PBAT to be used in tissue engineering to reconstruct or replace tissue (such as bone, cartilage, and blood vessels). On the other hand, reinforcing fillers must strengthen the PBAT matrix's performance and enhance its mechanical and thermal properties for broader use. Countless loadings in the PBAT matrices have been studied to improve the physical, thermal, permeability, and characteristics of the finished goods.

Nano-silica is used in combination with PBAT/P(MMA-co-MA) blends as a nanofiller to improve the characteristics of materials owing to its low cost, high reproducibility, and suitability for large-scale production [28]. SiO_2 nanoparticle (NP) synthesis is a topic of interest in basic and applied research. Other variables include the particle-size suitability for multiple products. The last few years have seen a significant increase in interest in SiO_2-filled composites. According to studies, composite materials loaded with SiO_2 possess good thermal and mechanical characteristics [29,30]. Studies have been carried out on the SiO_2 distribution in poly(methyl methacrylate) matrices [31,32], high-density polyethylene [33], and poly(ethylene oxide) [34]. PBAT is a thermoplastic compostable material commonly used to manufacture films, bottles, and fibers for different applications because of its outstanding mechanical and optical characteristics, susceptibility to fatigue and wear, and creep-fragmentation resistance [35,36]. Functional characteristics, such as antimicrobial activity, have been enhanced during the fabrication of PBAT-based composite films [37–39].

Recent research has been conducted on biomass or sustainable polymers reinforced with fillers and fabricated to use the solution casting method [40–42]. In this work, a series of PBAT/P(MMA-co-MA) composites with different contents was prepared by solution casting of PBAT/P(MMA-co-MA) blends and a SiO_2 nanopowder [43]. SiO_2 has a disk structure according to scanning electron microscopy (SEM). The excellent compatibility and interactions between SiO_2 and PBAT mean that SiO_2 NPs are dispersed uniformly in PBAT/P(MMA-co-MA) blends according to the shape of the film surface [44]. The tensile strength of the PBAT/P(MMA-co-MA) composites increased from 9.43 MPa for PBAT without reinforcement to 22.82 MPa with 5.0 wt.% of SiO_2 NPs (PBMS-3). Thermogravimetric analysis (TGA) of the composite films of PBAT/P(MMA-co-MA)–SiO_2 showed an increase in the weight loss temperature (%). The PBAT/P(MMA-co-MA)–SiO_2 composite materials also have high hydrophobicity. The antimicrobial activity of SiO_2-incorporated PBAT/P(MMA-co-MA) composites is remarkable because of the combined antimicrobial activity of SiO_2 NPs against *S. aureus* and *E. coli*. As a result, composites composed of PBAT/P(MMA-co-MA)–SiO_2 were developed and assessed for food packaging applications.

2. Materials and Methods

2.1. Chemicals and Materials

Poly(butylene adipate-co-terephthalate) (PBAT) pellets were supplied by BASF. The melt flow index (MFI) (190 °C, 2.16 kg) was 3.3–6.6 g/10 min. A solution of methyl methacrylate (MMA) was purchased from Daejung Chemicals in Korea. Maleic anhydride (MA; 98%, Sigma-Aldrich, St. Louis, MO, USA) was recrystallized from chloroform and n-hexane. By crystallizing from methanol, the radical initiator of azobisisobutyronitrile (AIBN; 99%, Sigma–Aldrich, St. Louis, MO, USA) was purified. Tetraethoxysilane (TEOS) was received from Sigma–Aldrich in India. Daejung Chemicals in Korea supplied methyl benzene, ethyl acetate, ethanol, and N, N-dimethylformamide (DMF). All compounds were used as received. For the preparation of solutions and reactions, double-distilled water was used.

2.2. Synthesis of SiO_2 Nanoparticles (NPs)

SiO_2 NPs were synthesized using the methodology reported elsewhere [45]. Tetraethoxysilane (TEOS) was used as a silica source in the sol–gel technique to prepare SiO_2 NPs. The reaction was rapidly stirred for 30 min at 150 °C with 7.4 mL TEOS and 80 mL methanol introduced. A white precipitate was heated for one hour at 350 °C in a tubular furnace to generate SiO_2. The XRD, SEM, and TEM characterization data matched the information from the previous report [46].

2.3. Synthesis of Poly(methyl methacrylate-co-maleic anhydride)

The methyl methacrylate and maleic anhydride monomers were prepared by a radical reaction to produce the P(MMA-co-MA) copolymer. Figure 1 depicts the synthesis of P(MMA-co-MA). Before use, the MMA monomer (1.50 g; 0.015 mM) underwent two reduced-pressure vacuum distillations at 35 °C. For 45 min, the polymerization reaction took place in a reactor with rapid mixing and a nitrogen atmosphere. The copolymers were synthesized using 3.43 g: 0.035 mM of (maleic anhydride) MA, ethyl acetate as the solvent, with AIBN as the radical polymerization initiator at 85 ± 1 °C. The method of synthesis used in this work can be found in the literature [12,47].

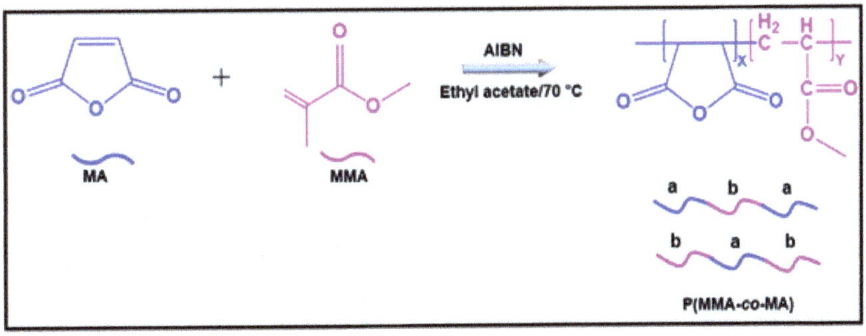

Figure 1. Synthetic routes of poly(methyl methacrylate-*co*-maleic anhydride) copolymer.

2.4. Purification of P(MMA-co-MA) Copolymer

The material was produced by mixing un-reacted MMA, MA, and P(MMA-*co*-MA) copolymer. After mixing the materials, the diluted solution was added to the diethyl ether in the precipitate for P(MMA-*co*-MA). A tetrahydrofuran/water (9/1 *v/v*) solution was used to dissolve the P(MMA-*co*-MA), resulting in a solution with a 2.0×10^{-2} g mL^{-1} specific concentration. The reaction was quenched when the solution became too viscous to be stirred (usually around 1 h). The reaction mixture was diluted in THF and precipitated in n-hexane to remove the unreacted initiator and monomer. This procedure was repeated three times to obtain the purified copolymer. The obtained P(MMA-*co*-MA) precipitate was dried under vacuum at 100 °C for 10 h.

2.5. Fabrication of PBAT/P(MMA-co-MA)–SiO$_2$ Composites

PBAT and P(MMA-*co*-MA) were blended at a 70:30 ratio. This ratio was selected as it also generated a suitable blend of properties because PBAT was the target polymer and its properties needed to be enhanced. Solution casting methods were used to fabricate the PBAT/P(MMA-*co*-MA)–SiO$_2$ composites [48]. A homogeneous solution was produced first by dissolving PBAT (1.0 g) in chloroform (50 mL) at room temperature for 12 h. Furthermore, 50 mL of P(MMA-*co*-MA) and chloroform (30 wt.% with respect to PBAT) were blended for one hour in an ultrasonic mixer. The produced solution was then transferred to a glass Petri dish after being sonicated, in which the chloroform solvent was allowed to dry completely. The chloroform was removed by scraping the PBAT/P(MMA-*co*-MA) composites off the Petri plate and drying them at 60 °C under vacuum for eight hours. The PBAT/P(MMA-*co*-MA) blends were stored in an airtight container for future analysis. In contrast, a PBAT and P(MMA-*co*-MA) film was prepared by adding chloroform solvent and casting it onto a glass Petri dish after mixing for an hour. Removing the dry films allowed samples to be kept in a desiccator at 23 °C until tested. On the premise of the wt.% of SiO$_2$ it includes, the PBAT/P(MMA-*co*-MA) blend film was given the designations PBMS-1, PBMS-2, and PBMS-3. Table 1 lists the data collected.

Table 1. Material formulation in blends and composite film preparation.

Samples	PBAT (wt.%)	P(MMA-*co*-MA) (wt.%)	SiO$_2$ NPs (wt.%)
PBAT	100.0	-	-
P(MMA-*co*-MA)	-	100.0	-
PBAT/P(MMA-*co*-MA)	70.0	30.0	-
PBMS-1	70.0	30.0	1.0
PBMS-2	70.0	30.0	3.0
PBMS-3	70.0	30.0	5.0

Abbreviations: PBAT, poly(butylene adipate-*co*-terephthalate); P(MMA-*co*-MA), Poly(methyl methacrylate-*co*-maleic anhydride); PBMS, PBAT/P(MMA-*co*-MA)–SiO$_2$; SiO$_2$, silicon dioxide.

2.6. Characterization

2.6.1. Structural Characterization

The ^{1}H-nuclear magnetic resonance (NMR) and ^{13}C-NMR (600 MHz) spectra were recorded on a Bruker instrument (OXFORD, AS600, USA) using deuterated chloroform (CDCl3) as the solvent and tetramethyl silane as the reference. The Fourier-transform infrared (FTIR, Perkin–Elmer Spectrum Two) spectra were obtained to produce the attenuated total reflection (ATR)-FTIR spectra in a 4000–400 cm^{-1} wavenumber range, and X-ray diffraction (XRD, Rigaku, PANALYTICAL) was performed at a scan rate of 0.50 min^{-1} over a scan range of 10 to 80° 2θ.

2.6.2. Morphological Studies

The surface morphology of clean PBAT, P(MMA-co-MA), and its composites was examined by SEM (JEOL 6400, Tokyo, Japan). TEM (JEOL, JEM-2100, Japan) was employed to investigate the inner microstructure of PBAT/P(MMA-co-MA) and composites. A small amount of the samples was sonicated in ethanol for descriptive purposes, and a drop of it was cast in a 300-mesh copper grid with carbon coating for electromagnetic measurements.

2.6.3. Thermal Characterization

Thermogravimetric–differential thermal analysis (TG–DTA, TA Instruments, SDT Q6000) was used to evaluate the thermal stability. Under a N_2 atmosphere, all TG–DTA was conducted from 50 °C to 700 °C at a scanning rate of 20 °C/min (nitrogen flow rate was 60 mL min^{-1}). Differential scanning calorimetry (DSC, A Instruments, DSC Q200). was performed using 5.0 mg of the PBAT/P(MMA-co-MA)–SiO_2 composite samples. The samples were heated from room temperature to 300 °C in a N_2 atmosphere at a flow rate of 50 mL per minute. The thermal history in the case of PBAT was removed by heating the sample to 180 °C and holding it at that temperature for two minutes. The sample was then cooled to 50 °C and heated to 180 °C again. The heating rate used for all DSC runs was 20 °C/min.

2.6.4. Mechanical Strength Measurements

The tensile strength of PBAT/P(MMA-co-MA) and its composites was evaluated on a universal testing machine (3345, Instron, Norwood, MA, USA) in accordance with ASTM-D882 at 23 °C and 50% RH. The PBAT composite samples used for the evaluation had the following dimensions: 50 × 20 mm, gauge length of 30 mm, and speed of 20 mm/min. Five tensile-strength samples were tested and the average result was calculated. In MPa, the maximum tensile strength was specified. The digital thickness measurement instrument (Mitutoyo micrometer, Japan) determined the film thickness to the closest 0.001 mm. The mean of at least five locations was used to determine the values. The mechanical and physical characteristics of the materials were measured to use the estimated values.

2.6.5. Barrier Properties

The OTR measurements for the PBAT/P(MMA-co-MA) and its composites were evaluated using a *Noselab* (ATS, Concorezzo, Italy) at 23 °C and 50% RH according to the ASTM D3985 standard method. The machine had a one-atm pressure. Five different locations on the composite samples were measured and the average value was used. The specimen was processed at room temperature. A *Lyssy L80-5000* was used to measure the WVTR values of the PBAT/P(MMA-co-MA) blends and PBAT/P(MMA-co-MA)–SiO_2 composites in accordance with ASTM F1249-90 under the conditions of 100% RH at 23 °C. Five repeats of the test were carried out prior to calculating the average result.

2.6.6. Water Contact Angle Measurements

Using a contact angle meter (Dataphysics Instruments, OCA-20, Filderstadt, Germany), the sessile drop method, at 23 °C, and 50% RH, the water contact angle of the PBAT and its composites was studied. A 1µL droplet of water was placed on PBAT and the composite surface of the films and a droplet image was taken within 5 s to measure the contact angle. The mean value was calculated after the surface tension measurements were performed at five different locations on the film. The experimental result has a $\pm 1°$ confidence interval.

2.6.7. Antimicrobial Activities

The antimicrobial activities of PBAT/P(MMA-co-MA) with SiO_2 NPs were tested using the zone-of-inhibition technique. In compliance with ASTM E 2149-01, the composites were tested in advancing contact conditions against Gram-negative (*E. coli*) and Gram-positive (*S. aureus*) microorganisms. To prepare a broth, the beef extract (1.0 g) and peptone were mixed in 100 mL of water (2.0 g). A shaking incubator at 40 °C and 200 rpm was applied to cultivate the solution for 24 h. The 0.90% sterile NaCl aqueous solution was used to dilute the microorganism cell suspension by a frequency of 10^6 for *E. coli* and *S. aureus*.

2.6.8. Statistical Analysis

ANOVA in SPSS 21 was used (IBM, New York, NY, USA) to determine the statistical significance of each result. The data are given as the mean ± standard deviation. Statistical differences were analyzed using a one-way analysis of variance, and a value of $p < 0.05$ was considered significant.

3. Results and Discussion

3.1. Characterization of SiO_2 Nanoparticles

FTIR spectroscopy of the synthesized SiO_2 NPs was conducted (Figure 2a). The two major peaks at 1077 and 460 cm^{-1} suggest the presence of SiO_2 NPs as asymmetric and symmetric Si-O-Si modes, respectively [49]. The Si–O stretching vibration for surface Si–OH groups was observed at 795 cm^{-1}. Rahman et al. suggested that the Si–O–Si and Si–OH groups on SiO_2 NPs help compensate for the silica network [50]. Figure 2b shows the XRD pattern of the SiO_2 NPs. Two reflection peaks were observed at 21.7° 2θ, which are at the scattering angle from the (101) lattice planes, showing that the SiO_2 is crystalline [51]. The Scherrer equation showed that the size of crystalline SiO_2 NPs was 20 nm. SEM and TEM were used to characterize the SiO_2 NP structure. SEM revealed the surface of SiO_2 (Figure 2c). The SiO_2 nanoparticles had a mean diameter of 20 nm and a uniform morphology [52]. SiO_2 ranged in size from 20 to 50 nm, and TEM showed that it was almost spherical (Figure 2d). Figure 2d's inset shows the selected area electron diffraction (SAED) pattern of synthesized SiO_2. Only a few diffraction spots scattered in a circle can be seen when electron diffraction is conducted on a small proportion of crystals. The (101) and (222) planes of the face-centered cubic silica structure were fitted by the patterns. It is also feasible to determine the crystalline behavior from the SAED image.

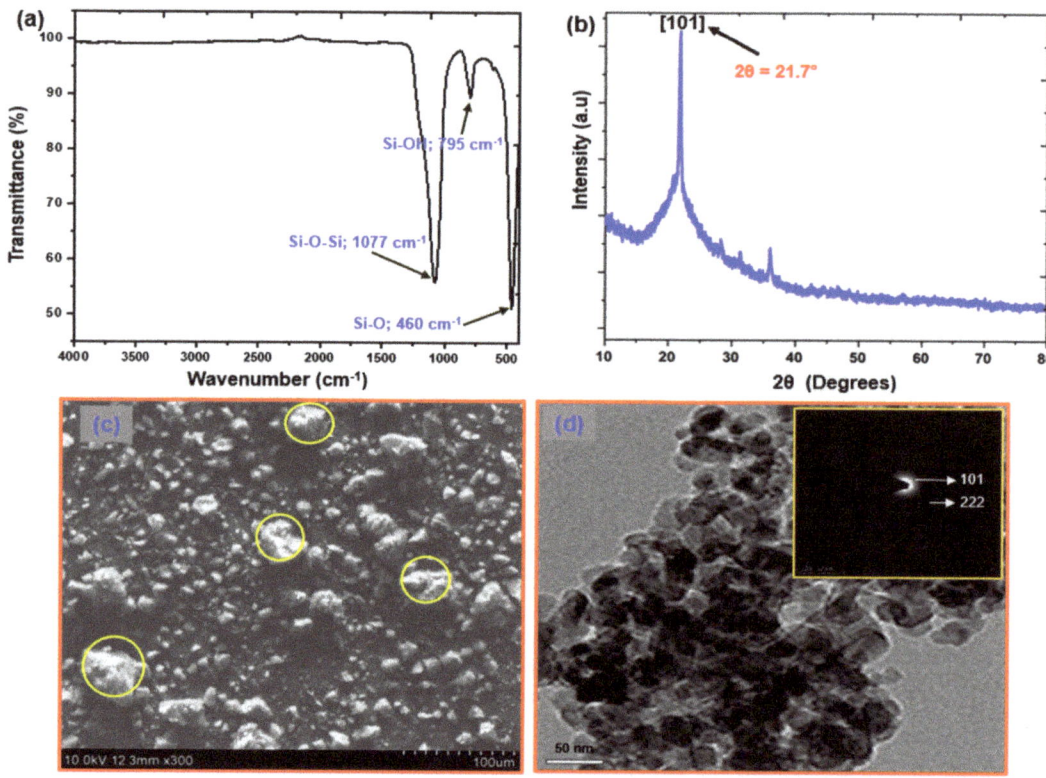

Figure 2. Structural and morphological characterization of SiO$_2$ nanoparticles: (**a**) FTIR spectrum; (**b**) XRD pattern; (**c**) SEM; and (**d**) TEM images (inset in d is the SAED pattern).

3.2. Characterization of P(MMA-co-MA) Copolymer

Figure 3a presents the P(MMA-*co*-MA) and ^{1}H-NMR spectra. The chemical shifts at 5.55 ppm and 6.09 ppm in the copolymers were produced by the double-bond proton of the MMA molecule. The copolymer MA units, which have methylene protons, are responsible for the absorption peaks at 6.35 and 7.25 ppm. The integrated areas at 0.65 and 1.45 ppm reflect the methyl protons of the MMA units and the integrated areas at 1.65 and 2.07 ppm reflect the methylene protons [53]. The methyl protons of –COOCH$_3$ are essential for the absorption peaks at 3.42 and 3.87 ppm. Figure 3b displays the ^{13}C-NMR spectrum of the P(MMA-*co*-MA) copolymer. From 180.7 to 182.2 ppm, the carbonyl carbon (>C=O) signals of both MA and MMA units could be seen. The back and side chain of the P(MMA-*co*-MA) copolymer show aliphatic carbon resonance in a spectral region from 26.9 to 38.0. In 62.2 (^{6}CH$_2$), 43.7 (^{4}CH$_2$), and 38.2, the side-chain-ring methylene carbon signal was assigned (^{5}CH$_2$). The methyl carbon of MMA is shown by the signal at 45.9 (^{11}CH$_3$). The carbon (^{1}CH) and CH$_2$ signals overlap.

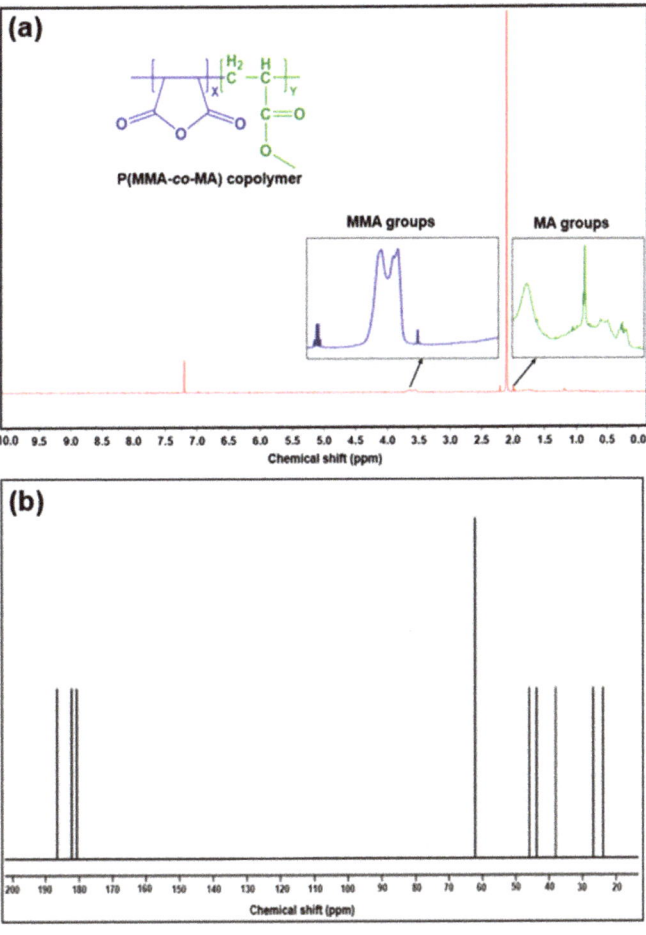

Figure 3. (a) ^{1}H-NMR spectrum; (b) ^{13}C-NMR spectrum of the P(MMA-co-MA) copolymer.

The FTIR spectra were used to study the P(MMA-*co*-MA) copolymer structures, as shown in Figure 4a. The asymmetric and symmetric stretching modes of C=O are represented by all of these anhydride units (1778 and 1850 cm^{-1}), which are also found in P(MMA-*co*-MA). In addition, the distinct anhydride band (O=C–O–C=O; 950 cm^{-1}) indicated that MA had been successfully introduced into the copolymers. The stretching vibration of the ester group (O–C=O; 1718 cm^{-1}) in MMA was observed. The carbonyl groups of MA and MMA show three distinct peaks that closely follow one another, as indicated. Moreover, the maximum of the C–H stretching mode of MMA and MA (2948 cm^{-1}), in addition to symmetric and asymmetry bend vibration of the -CH$_3$ and -OCH$_3$ bridges of the MMA unit (1436 and 1390 cm^{-1}, accordingly), all indicated the existence of these monomer units within the resulting copolymers. MMA and MA performed a copolymerization reaction at 85 °C. Figure 4b shows the XRD patterns of synthetic P(MMA-*co*-MA) copolymer. XRD is the most outstanding method for smoothly finishing the material structure. The XRD pattern of P(MMA-*co*-MA) [54]. The characteristic broad peak at 19.5° 2θ, corresponding to the (111) plane, reveals the amorphous of P(MMA-*co*-MA). Figure 4c shows the SEM images of the synthetic P(MMA-*co*-MA) copolymer. Although the P(MMA-*co*-MA) SEM images are homogeneous and smooth, the P(MMA-*co*-MA) exhibits aggregated molecular structures. TEM was used to study the inner morphology of synthesized P(MMA-*co*-MA) copolymer. Figure 4d shows TEM images of the P(MMA-co-MA) matrix. The TEM images revealed a transparent portion of the images, representing the smooth and uniform structure of the P(MMA-co-MA). Figure 4(d)'s inset depicts the selected area electron diffraction (SAED) pattern of the synthesized P(MMA-co-MA) copolymer. The amorphous structure of the P(MMA-co-MA) copolymer is evident in its SAED image.

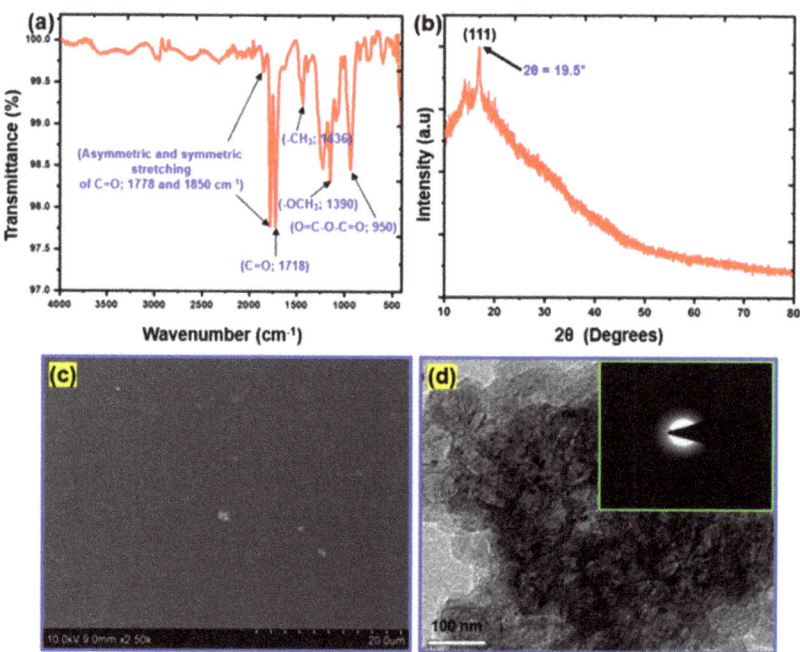

Figure 4. Structural and morphological characterizations of P(MMA-co-MA) copolymer: (**a**) FTIR spectrum; (**b**) XRD pattern; (**c**) SEM; and (**d**) TEM images (inset in d is the SAED pattern).

3.3. Characterization of PBAT/P(MMA-co-MA)–SiO₂ Composites

Figure 5A shows the FT-IR spectra. The distinct peak at 2958 cm^{-1} was assigned to the stretching vibration of C–H groups [55,56]. The prominent peaks at 1780 and 1720 cm^{-1} were produced by the strong peak of the C=O group in the PBAT and P(MMA-co-MA). The peaks suggested the ester linkage (C–O) at 1270 and 1110 cm^{-1} of P(MMA-co-MA) copolymer. The bending vibration of the C–C can be ascribed to the peak at 1450 and 1390 cm^{-1}, respectively. The vibration of the adjacent (CH$_2$–) groups of the PBAT matrix was observed at 720 cm^{-1}. In SiO$_2$, the presence of silicon is confirmed by the peaks at 475 and 850 cm^{-1}. In the PBAT/P(MMA-co-MA)–SiO$_2$ ternary composites, the peaks for the PBAT/P(MMA-co-MA) blend film moved toward higher and lower wavenumbers. The shift toward a higher wavenumber is due to the link between the –COO group in PBAT and SiO$_2$ through metal bonding. The FTIR results showed SiO$_2$ with a suitable molecular attachment in the PBAT/P(MMA-co-MA) blends. The semi-crystalline characteristics of the PBAT/P(MMA-co-MA)–SiO$_2$ composites were examined with XRD. Figure 5B shows XRD patterns of PBAT/P(MMA-co-MA)–SiO$_2$ composites, and the prominent peaks were observed at 15.7°, 17.8°, 20.4°, 21.8°, 23.2°, 25.4°, and 28.0° 2θ, corresponding to lattice planes of (010), (020), (012), (110), (102), (210), and (101), respectively. PBAT/P(MMA-co-MA)–SiO$_2$ composite films show two more peaks at 21.7° due to inorganic SiO$_2$ NPs on the film surface [54]. The neat PBAT/P(MMA-co-MA) blend exhibits reflection planes of (102), (012), and (113) for PBAT and the (002) peak for P(MMA-co-MA), respectively. Therefore, the PBAT/P(MMA-co-MA) blend forms a heterogenous phase because of the poor miscibility or weak interactions between the two composites. These three peaks are found in the same positions throughout all PBAT/P(MMA-*co*-MA) blends, indicating that the SiO_2 NPs alter the semi-crystal form of the PBAT/P(MMA-*co*-MA). These results also suggest that adding SiO_2 as a filler has no noticeable effect on the semi-crystal form of PBAT/P(MMA-*co*-MA). The clean PBAT, P(MMA-*co*-MA) film, and PBAT/P(MMA-*co*-MA)–SiO_2 composites with different percentages of (1.0 to 5.0 wt.%) SiO_2 showed uniform surfaces because the viscosities of the respective film-forming solutions were suitable for casting films (Figure 5C). The PBAT/P(MMA-*co*-MA)–SiO_2 composite film with 5.0 wt.% SiO_2 (PBMS-3), however, had a rougher surface because of SiO_2 agglomeration and air bubbles trapped in the casting solution. These flaws occurred because the viscosity of the PBAT solution with 5.0 wt.% SiO_2 prevented air bubbles from exiting and the SiO_2 content above the optimal value (~3.0 wt.%) enhanced PBAT and metal oxide (SiO_2) interactions. Therefore, using SiO_2 NPs, the effects of SiO_2 on the barrier and mechanical characteristics of the PBAT/P(MMA-*co*-MA) blends were evaluated. The results were compared with the outcomes of the PBAT and PBAT/P(MMA-*co*-MA) composites.

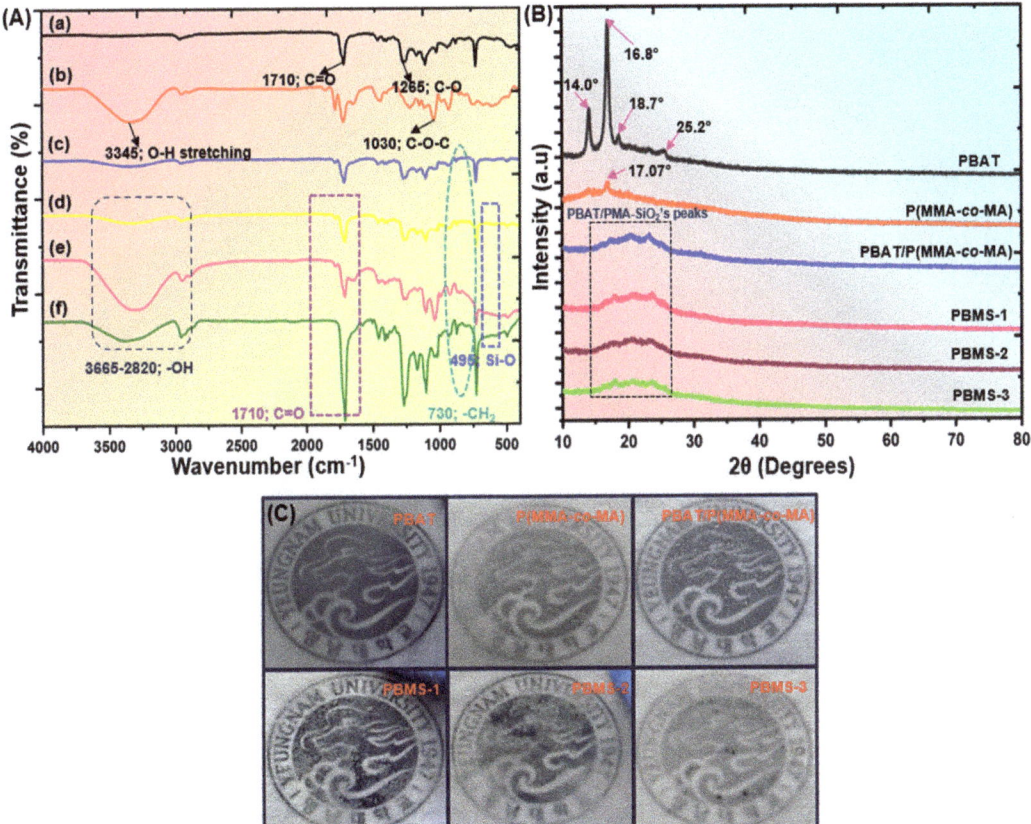

Figure 5. Structural characterization of PBAT/P(MMA-*co*-MA)–SiO_2 composites: (**A**) ATR-FTIR spectra ((a) PBAT; (b) P(MMA-*co*-MA); (c) PBAT/P(MMA-*co*-MA) blends; (d) PBMS-1; (e) PSMS-2; (f) PBMS-3); (**B**) XRD patterns; (**C**) the appearance of the film.

Figure 6 presents SEM images of the PBAT/P(MMA-*co*-MA)–SiO_2 composites prepared with different wt.% loadings (SiO_2). The figures show that the presence of SiO_2 NPs in PBAT/P(MMA-*co*-MA) blends showed that the PBAT and P(MMA-*co*-MA) have smooth surfaces that face downward and that a rough surface was observed after loading a larger weight % of SiO$_2$ NPs. Significant SiO$_2$ aggregates dispersed in the PBAT/P(MMA-*co*-MA) were observed in the SEM images, as illustrated by the red circles in Figure 6d. The SiO$_2$ and the PBAT/P(MMA-*co*-MA) blend were visible, showing the poor compatibility of the two different polymers. Because of the similar composition ratios, PBAT/P(MMA-*co*-MA) (PBM-3) formed a continuous structure, as shown in Figure 6f. The green arrow in Figure 6e also shows that the formation of substantial amounts of SiO$_2$ was induced by the ultra PBAT upon impact splitting. This played a major role in the high tensile strength of the ternary composites. The SEM image of the PBAT/P(MMA-*co*-MA)–SiO$_2$ composite film was produced using the lowest amount of SiO$_2$ (1.0 wt.%) (PBMS-1). In this image, the SiO$_2$ in the PBAT/P(MMA-*co*-MA) blend is distributed evenly. SiO$_2$ NPs agglomerate in the PBAT/P(MMA-*co*-MA) matrix as a result of the increased interactions among nanoparticles as the SiO$_2$ concentration is increased. Based on these results, the C=O units of PBAT and P(MMA-*co*-MA) and SiO$_2$ have strong bonding interactions in the developed ternary PBAT/P(MMA-*co*-MA)–SiO$_2$ composites. SEM was used to establish that the SiO$_2$ NPs and PBAT/P(MMA-*co*-MA) blends constituted a nanostructured composite material.

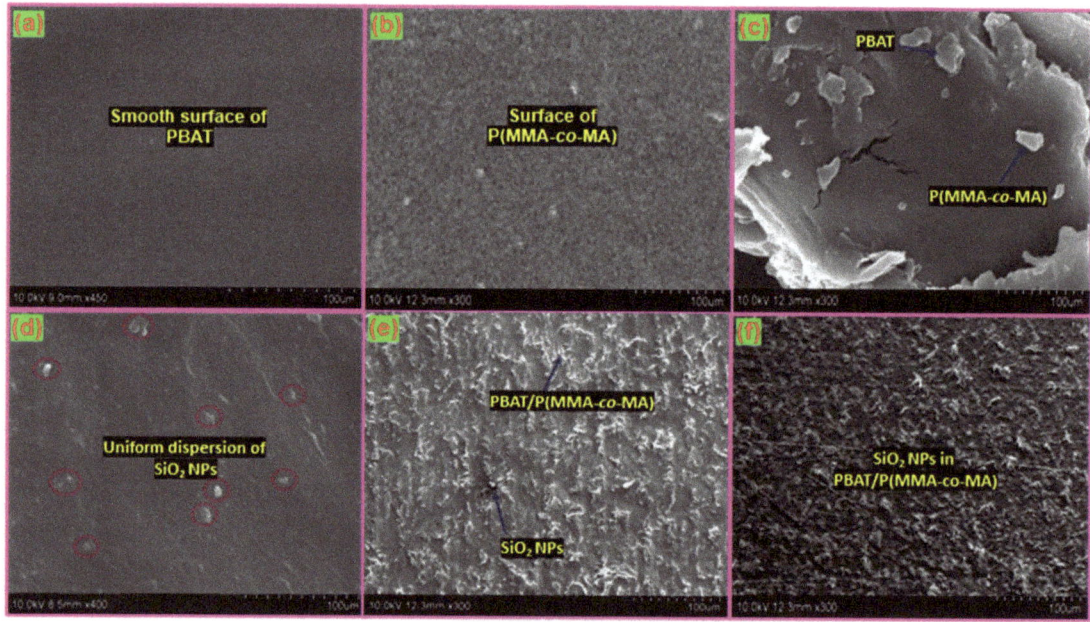

Figure 6. SEM images of the surfaces of PBAT/P(MMA-*co*-MA) blends and their composites: (**a**) PBAT, (**b**) P(MMA-*co*-MA), (**c**) PBAT/P(MMA-*co*-MA) blends, (**d**) 1.0, (**e**) 3.0, and (**f**) 5.0 wt.% of SiO$_2$ NPs.

The inner morphology of the PBAT/P(MMA-*co*-MA)–SiO$_2$ composites was investigated using TEM and the results are shown in Figure 7. The SiO$_2$ NPs were encapsulated in the PBAT/P(MMA-*co*-MA) matrix, as shown in these images. The transparent region of the images is a representation of the PBAT/P(MMA-*co*-MA) blend matrix. Figure 7a,b show the inner morphologies of PBAT and P(MMA-*co*-MA), and Figure 7c–e show a TEM image of PBAT/P(MMA-*co*-MA) blends, PBMS-1, PBMS-2, and PBMS-3 composites. The PBAT/P(MMA-*co*-MA) blend shows uneven morphology (small round structure and agglomerate merged) owing to the poor miscibility in the polymer blend, as seen in Figure 7c. The inner morphology of PBAT/P(MMA-*co*-MA)–SiO$_2$ composites (Figure 7d–f) showed a distorted and smooth morphology owing to the bonding strength of the polymer matrix and SiO$_2$ surface. Agglomerations of SiO$_2$ NPs in the PBAT/P(MMA-*co*-MA) matrix are seen in Figure 7f. SiO$_2$ had a high dispersion in the matrix of the PBAT/P(MMA-co-MA) blend. Independent of the SiO$_2$ concentration, all the composites showed a dispersed morphology of SiO$_2$ when evaluating how the SiO$_2$ of the PBAT and P(MMA-co-MA) influences the morphology of the systems. The composites of PBAT/P(MMA-co-MA)–SiO$_2$ showed fewer interactions and inner morphology according to the TEM of PBMS-3. SiO$_2$ is a suitable reinforcing filler and compatibilizer in PBAT/P(MMA-co-MA) blends, as shown by TEM.

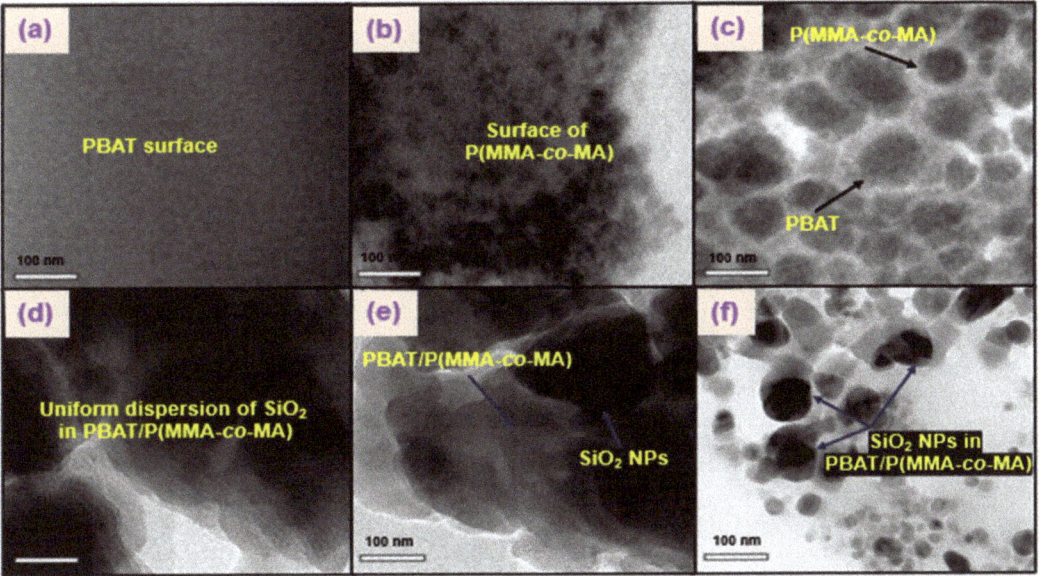

Figure 7. TEM images of PBAT/P(MMA-co-MA)–SiO$_2$ ternary composites: (**a**) PBAT, (**b**) P(MMA-co-MA), (**c**) PBAT/P(MMA-co-MA) blends, (**d**) PBMS-2, (**e**) PBMS-2, and (**f**) PBMS-3.

3.4. Thermal Properties of PBAT/P(MMA-co-MA)–SiO$_2$ Composite Films

3.4.1. Thermogravimetric Analysis (TGA)

Figure 8A shows the TGA curves of PBAT/P(MMA-co-MA) composites with different SiO$_2$ NPs concentrations. Table 2 lists the thermal parameters. The two-step thermal degradation behavior was observed in all PBAT/P(MMA-co-MA)–SiO$_2$ composites. The weight loss at 350 and 450 °C was attributed to the thermal degradation of P(MMA-co-MA). The weight loss at temperatures above 342.3 °C was caused mostly by the degradation of PBAT. The least stable blend of PBAT and P(MMA-co-MA) was without SiO$_2$ NPs. The temperature of the initial weight loss of the PBAT/P(MMA-co-MA) blend was only 382.5 °C, and the temperature of the final weight loss was only 403.1 °C. The thermal stability of the composites improved significantly as the SiO$_2$ NPs content increased. SiO$_2$ could act as a barrier inside the composites, slowing the diffusion of the decomposition of PBAT and P(MMA-co-MA) while limiting oxidative degradation and increasing the thermal stability of the composite materials. PBAT/P(MMA-co-MA)–SiO$_2$ (PBMS-3) composites showed the highest thermal stability of the PBAT/P(MMA-co-MA) composites containing different amounts of SiO$_2$, whereas PBAT/P(MMA-co-MA)–SiO$_2$ (PBMS-1) exhibited the lowest thermal stability. SiO$_2$ can graft onto polymer chains owing to the ability of SiO$_2$ NPs to interact with P(MMA-co-MA) or PBAT. These interactions improve their compatibility and dispersibility in polymer matrices and their physical barrier effect within the matrix. Another factor is the effectiveness of the reaction between the PBAT/P(MMA-co-MA) and SiO$_2$. The PBAT/P(MMA-co-MA)–SiO$_2$ (PBMS-3) composites showed higher thermal stability.

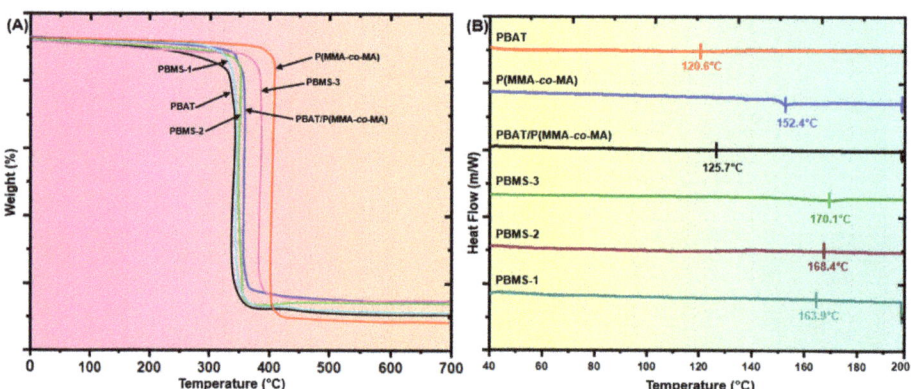

Figure 8. (**A**) TGA and (**B**) DSC curves of PBAT, P(MMA-*co*-MA), PBAT/P(MMA-*co*-MA), and PBAT/P(MMA-*co*-MA)–SiO$_2$ composites.

Table 2. TGA and DSC results of PBAT/P(MMA-*co*-MA) blends and their composites.

Samples	TGA			DSC	
	Initial Degradation Temperature (°C) [a]	Final Degradation Temperature (°C) [b]	Ash Content (%) [c]	T_g (°C)	T_m (°C)
PBAT	322.3	358.2	3.59	49.3	120.6
P(MMA-*co*-MA)	395.0	420.6	2.56	53.9	152.4
PBAT/P(MMA-*co*-MA)	368.4	385.2	1.68	56.0	125.7
PBMS-1	348.1	359.4	1.13	52.1	163.9
PBMS-2	364.9	379.0	1.41	60.7	168.4
PBMS-3	382.5	403.1	2.06	65.2	170.1

[a] Temperature at which the initial mass loss was recorded. [b] Temperature at which the final mass loss was recorded. [c] Mass percentage of material remaining after TGA at a maximum temperature of 700 °C.

3.4.2. Differential Scanning Calorimetry (DSC)

The melting temperature (Tm) and glass transition temperature (Tg) were measured for PBAT/P(MMA-*co*-MA)–SiO$_2$ composites using DSC to evaluate the thermal stability of film components. Figure 8B shows the DSC thermograms of PBAT/P(MMA-*co*-MA) and PBAT/P(MMA-*co*-MA)–SiO$_2$ composites. Table 2 lists the relevant DSC data. The PBAT and P(MMA-*co*-MA) film had a T_g of 49.3 and 53.9 °C, respectively. In contrast, the SiO$_2$-induced PBAT/P(MMA-*co*-MA) composites showed a T_g (52.1 to 65.2 °C) with an increase in SiO$_2$ content, indicating good compatibility of the components of the (PBMS) composites. Furthermore, the melting temperature (Tm) of the PBAT/P(MMA-*co*-MA) blend film was 125.7 °C, and the PBAT/P(MMA-*co*-MA)–SiO$_2$ composites showed a single melting temperature (Tm) that increased from 163.9 to 170.1 °C as the SiO$_2$ content was increased from 1.0 to 5.0 wt.%. The T_g and Tm values in the composites of PBAT/P(MMA-*co*-MA)–SiO$_2$ suggest that the ternary components interact to generate good comparability among PBAT, P(MMA-*co*-MA), and SiO$_2$. The melting temperature values of composites increased growth in the SiO$_2$ from 1.0 to 5.0 wt.%, showing that SiO$_2$ incorporation enhanced PBAT crystallization. The addition of 1.0 wt.% SiO$_2$ reduced the crystallinity of the PBMS-1 film significantly compared to the PBAT/P(MMA-*co*-MA) blend. By contrast, the addition of 3.0 and 5.0 wt.% SiO$_2$ significantly increased the crystallinity of the composite films PBMS-2 and PBMS-3.

3.5. Mechanical Strength of PBAT/P(MMA-co-MA)–SiO$_2$ Composite Films

Inorganic nanofillers are frequently incorporated into polymers to strengthen the mechanical characteristics of the resulting composites. In particular, the materials used to package food should be strong and rigid enough to support themselves and resist handling damage. The tensile and elongation at break (EB) of PBAT/P(MMA-co-MA)–SiO$_2$ ternary composites are higher than those of PBAT and PBAT/P(MMA-co-MA) blends, as shown in Figure 9. The tensile strength of the PBAT/P(MMA-co-MA)–SiO$_2$ ternary composites was SiO$_2$-dependent, as shown in Figure 9A. The tensile strength of the PBAT/P(MMA-co-MA)–SiO$_2$ composites increased as the SiO$_2$ concentration increased. This could be because of the strong bonding between the PBAT/P(MMA-co-MA) blends and SiO$_2$ interfacial adhesion. The elongation at break of the ternary composites decreased from 394.28 to 230.12% as the SiO$_2$ loading increased (Figure 9B). This is because SiO$_2$ is blended into PBAT/P(MMA-co-MA), decreasing the van der Waals strength among PBAT/P(MMA-co-MA) blend and SiO$_2$ NPs.

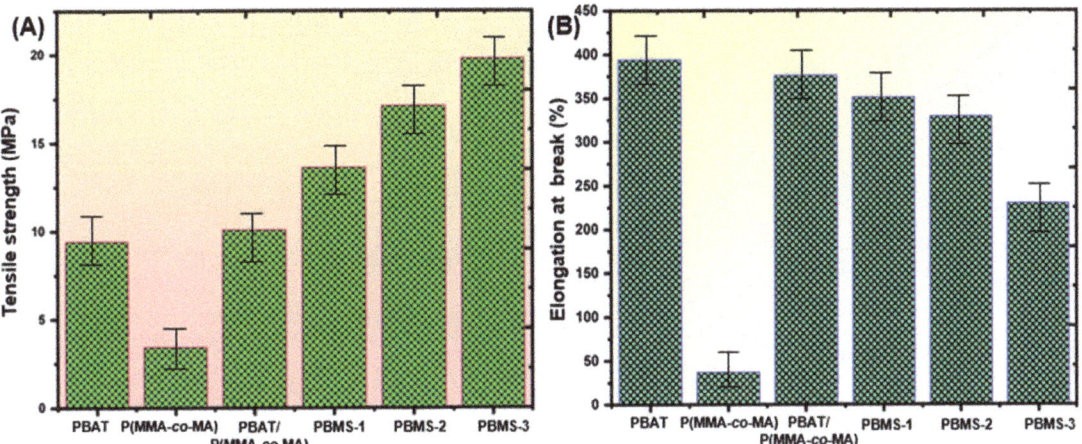

Figure 9. Mechanical characterizations of PBAT/P(MMA-co-MA)–SiO$_2$ composites: (**A**) tensile strength (MPa) and (**B**) elongation at break (%). Error bars represent ±5.01 standard errors.

The tensile and EB of neat PBAT matrix (9.43 MPa and 394.2%, respectively) and PBAT/P(MMA-co-MA) blends (10.10 MPa and 376.1%, respectively) were similar, resulting from less attraction to the PBAT/P(MMA-co-MA). On the other hand, a noteworthy improvement in tensile strength (19.81 MPa, a 230.1% increase) was facilitated by the high content of SiO$_2$ NPs (5.0 wt.%). The enhancement in tensile strength appears to result from an H-bond connection between the ester bond of the PBAT/P(MMA-co-MA) matrix and the Si–O group of the SiO$_2$ NPs. Because H-bonds and van der Waals forces are produced in PBAT/P(MMA-co-MA)–SiO$_2$ (PBMS-3) composites, their mechanical properties are improved, resulting in a strong interaction between them. The tensile and EB of PBMS-3 composites were superior to those of other PBMS films because of the surface function. With the addition of SiO$_2$, the lower percentage of the elongation at break from 34.28% to 18.01%, the brittle interaction between the SiO$_2$ and the PBAT/P(MMA-co-MA) blended material to become less flexible could be responsible.

3.6. Water Contact Angle

The water contact angles of the composites were measured. The film area appeared to be round because of the sudden change in the film after droplets were placed on the surface for every film in Figure 10. Figure 10 shows the pure PBAT film, revealing a contact angle of =69.4°, indicating its hydrophobicity. The pure PBAT film has a contact angle consistent with current research [55,56]. In contrast to P(MMA-co-MA), hydrophobicity was confirmed with a contact angle of =61.1°. The hydrophobic materials on the surface of the film of PBMS-3 have a contact angle of 86.2°. SiO$_2$ addition resulted in a higher contact angle than P(MMA-*co*-MA) (25.1°) and a lower angle than PBAT films (16.8°) after addition. A comparison of PBAT/P(MMA-*co*-MA) with the PBMS-3 film showed that the contact angle of these films was lower, and the affinity for water increased. As SiO_2 is incorporated into PBAT and P(MMA-*co*-MA) blend, the polar groups can give a contact angle that lowers the upward aspect.

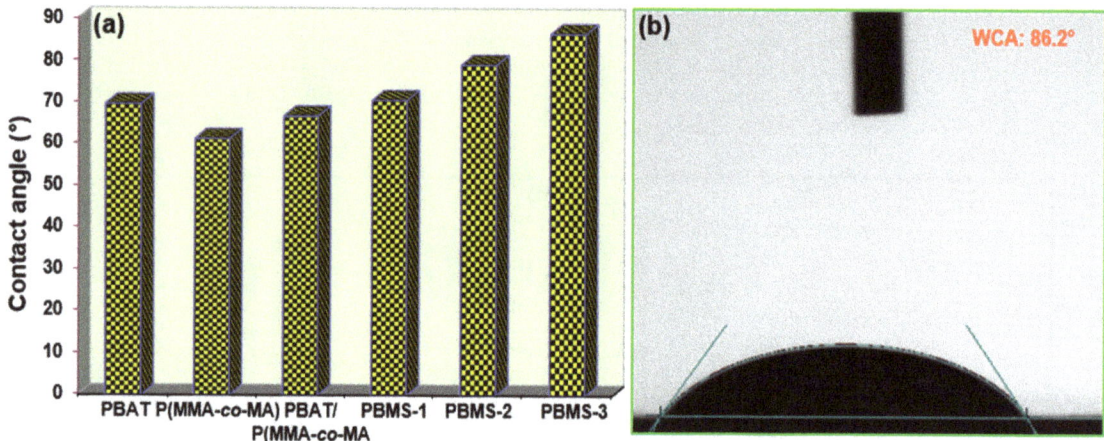

Figure 10. Contact angle of composite film samples (**a**) and PBAT/P(MMA-*co*-MA)–SiO_2 (5.0 wt.%) image (**b**).

3.7. Barrier Properties

Food packaging materials should have barrier properties for oxygen, or water may accelerate food spoilage in packing. It is essential to have low OTR and WVTR [57]. Table 3 lists the OTR and WVTR values of composite films produced by PBAT/P(MMA-*co*-MA)–SiO_2 with all these characteristics. The NP content caused a decrease in the OTR and WVTR values, indicating that the PBAT/P(MMA-*co*-MA)–SiO_2 ternary composites have good barrier properties. Previous studies reported activities in the OTR and WVTR for SiO_2-reinforced PBAT [58]. Water and oxygen molecules can travel through PBAT films completely free of impurities. On the other hand, these small molecules in the PBAT/P(MMA-*co*-MA)–SiO_2 composites encounter a long and circuitous pathway because of the requirement to migrate over or through the interfaces of impassable SiO_2.

Table 3. Barrier properties of PBAT/P(MMA-*co*-MA)–SiO_2 composite films.

Samples	Oxygen Transmission Rate, (cc m^{-2} Per 24 h)	Water Vapor Transmission Rate, (g m^{-2} Per 24 h)
PBAT	1137.2 ± 2.5 [a]	127.1 ± 2.9 [b]
P(MMA-*co*-MA)	860.2 ± 3.0 [a]	41.0 ± 2.2 [a]
PBAT/P(MMA-*co*-MA)	1095.6 ± 2.0 [c]	82.9 ± 3.1 [c]
PBMS-1	824.1 ± 3.4 [a]	63.2 ± 2.7 [c]
PBMS-2	589.3 ± 2.7 [b]	41.0 ± 3.0 [a]
PBMS-3	318.9 ± 2.0 [c]	26.3 ± 2.5 [a]

a–c: Different letters within the same column indicate significant differences among the film samples ($p < 0.05$).

As a result, the WVP and OTR values are decreased when SiO_2 NPs are introduced. WVTR must be reduced to reduce water transfer between the packing material and the material used for packaged foods. According to Table 3, the occurrence of SiO_2 NP content leads the oxygen transmission rate values ranging from 1137.2 to 318.9 (cc m^{-2} per 24 h). SiO_2 incorporation into the PBAT/P(MMA-*co*-MA) film reduced the OTR values of the ternary composites. The WVTR of the PBAT/P(MMA-co-MA) blend film was 82.9 (g m^{-2} per 24 h), whereas the WVTR of the PBAT/P(MMA-co-MA)–SiO$_2$ composites was lowered by the addition of SiO$_2$ NPs to 26.3 (g m^{-2} per 24 h). The WVTR of neat PBAT is 127.1 (g m^{-2} per 24 h). In the ternary PBAT/P(MMA-co-MA)–SiO$_2$ composites, there was a modest reduction in WVTR in the event of PBAT because of a decrease in hydrogen bonding with PBAT/P(MMA-co-MA) and the base film. This was attributed to the water-available absorption area. Incorporating 1.0 wt.% SiO$_2$ NPs significantly decreased the WVTR value to 63.2 (g m^{-2} per 24 h). The permeability decreased dramatically at the highest concentration of SiO$_2$ NPs (5.0 wt.%). SiO$_2$ NP loading decreased the WVTR by allowing water vapor to flow in zigzag pathways through the dispersed nanoparticles. On the other hand, the SiO$_2$ NPs tended to band together at higher concentrations, which decreased the effective content that promoted WVTR.

3.8. Antimicrobial Activities of PBAT/P(MMA-co-MA)–SiO$_2$ Composites

The antimicrobial properties of the metal oxide nanoparticles were good, and their reinforcement into the polymeric matrix significantly improved the antimicrobial property of the film. The zone-of-inhibition method was used to evaluate the antimicrobial property of PBAT/P(MMA-co-MA)–SiO$_2$ composite films. Figure 11 shows the results, and Table 4 lists the diameters of the film inhibition zones after calculating their specimen size. Th PBAT, P(MMA-co-MA), and PBAT/P(MMA-co-MA) films had no antimicrobial activities. The PBAT/P(MMA-co-MA) film loaded with a specific amount of SiO$_2$ will induce a zone of inhibition for microorganisms that are pathogenic to food. In tests against S. aureus and E. coli at concentrations of 1.0, 3.0, and 5.0 wt.% SiO$_2$, the PBAT/P(MMA-co-MA)–SiO$_2$ composites showed good antimicrobial activities compared to the PBAT and PBAT/P(MMA-co-MA) blends. The PBAT/P(MMA-co-MA)–SiO$_2$ ternary composites showed good antimicrobial characteristics when the SiO$_2$ NP concentration was as low as 1.0 wt.% (minimum (E. coli) 10.6 mm; (S. aureus) 9.2 mm), whereas the maximum ((E. coli) 17.9 mm; (S. aureus) 14.0 mm) was observed when the SiO$_2$ concentration was 5.0 wt.%. Compared to the PBAT composites, which are reported elsewhere, the antimicrobial properties were stronger [59]. In contrast to the report of a film incorporating SiO$_2$, the SiO$_2$ NP-incorporated high-antimicrobial-diameter film had a significantly higher zone of inhibition than the PBAT for the same composition. Increasing the SiO$_2$ concentration expanded the inhibition zone of the SiO$_2$-incorporated ternary composite films, showing that the PBAT/P(MMA-co-MA) film can function as a film that is active against both pathogens. E. coli was negatively charged and had less surface area than S. aureus. Furthermore, based on the observations, the PBMS-3 composite film exhibited strong antimicrobial activity against S. aureus and E. coli than the other films [60].

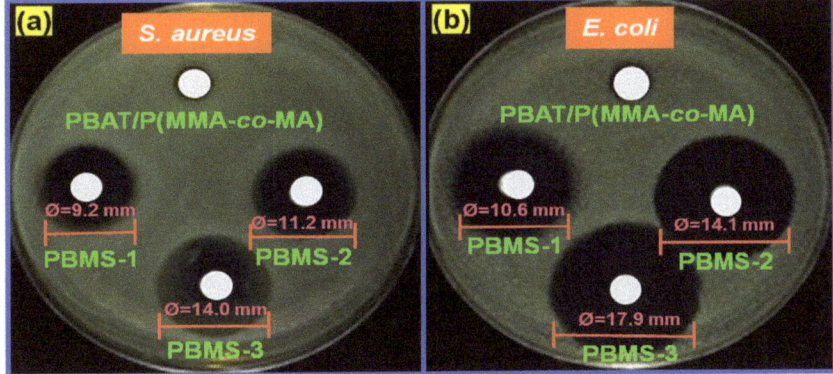

Figure 11. Antimicrobial test results of PBAT/P(MMA-co-MA) and their composite films against (**a**) S. aureus and (**b**) E. coli.

Table 4. Antimicrobial activity test values of the PBAT/P(MMA-co-MA) and PBAT/P(MMA-co-MA)–SiO$_2$ composites against S. aureus and E. coli.

Strain	Zone of Inhibition in (mm)			
	PBAT/P(MMA-co-MA)	PBMS-1	PBMS-2	PBMS-3
S. aureus	-	9.2 ± 3.84 [c]	11.2 ± 1.85 [a]	14.0 ± 2.63 [b]
E. coli	-	10.6 ± 2.35 [b]	14.1 ± 3.31 [c]	17.9 ± 2.56 [c]

Results are quoted as the mean ± standard deviation of three replicates. a–c: Different letters within the same column indicate significant differences among the film samples ($p < 0.05$).

4. Conclusions

A poly(methyl methacrylate-co-maleic anhydride) P(MMA-co-MA) copolymer was produced by radical polymerization. Solution casting was used to produce PBAT/P(MMA-co-MA) composite films with different SiO$_2$ concentrations. The addition of SiO$_2$ had a major effect on the mechanical, H$_2$O and O$_2$ barrier properties, thermal properties, and antimicrobial activity characteristics of the film. The tensile strength of the PBAT/P(MMA-co-MA) film was enhanced by the addition of SiO$_2$. The inclusion of SiO$_2$ improved the miscibility between PBAT and P(MMA-co-MA), according to the SEM and TEM results. The PBMS-3 composite film enhanced the elongation at break and tensile strength after adding SiO$_2$ to the PBAT/P(MMA-co-MA), allowing it to be studied as a structurally steady food packaging. Compared to PBAT/P(MMA-co-MA), the WVTR and OTR were much lower for the PBAT/P(MMA-co-MA)–SiO$_2$ composite films. Furthermore, SiO$_2$ produced PBAT/P(MMA-co-MA) materials that were more hydrophobic by increasing the contact angle. The water contact angle of PBAT/P(MMA-co-MA) was improved from 61.1 to 86.2 by introducing additional SiO$_2$ NPs, which enhanced the hydrophobicity. SiO$_2$-incorporated PBAT/P(MMA-co-MA) composite films showed effective antimicrobial activity against S. aureus and E. coli. The results suggest that the PBAT/P(MMA-co-MA)–SiO$_2$ composites can be used as materials for food packaging, which minimizes the microbiological load and extends the shelf life of packaged foods.

Author Contributions: Conceptualization, R.V. and K.A.; methodology, C.J.R.; software, C.J.R.; validation, V.R. and R.V.; formal analysis, R.V.; investigation, K.A.; resources, C.J.R.; data curation, D.S.; writing—original draft preparation, R.V.; writing—review and editing, D.S.; visualization, V.R.; supervision, S.-C.K.; project administration, S.-C.K.; funding acquisition, S.-C.K. All authors have read and agreed to the published version of the manuscript.

Funding: This research was supported by the Basic Science Research Program through the National Research Foundation of Korea (NRF), funded by the Ministry of Education (2020R1I1A3052258). In addition, the work was also supported by the Technology Development Program (S3060516), funded by the Ministry of SMEs and Startups (MSS, Republic of Korea), in 2021.

Institutional Review Board Statement: Not applicable.

Data Availability Statement: Not applicable.

Conflicts of Interest: The authors declare no conflict of interest.

References

1. Ramos, Ó.L.; Pereira, R.N.; Cerqueira, M.A.; Martins, J.R.; Teixeira, J.A.; Malcata, F.X.; Vicente, A.A. Bio-based nanocomposites for food packaging and their effect in food quality and safety. In *Food Packaging and Preservation*; Elsevier Inc.: Amsterdam, The Netherlands, 2018; pp. 271–306. [CrossRef]
2. Coppola, G.; Gaudio, M.T.; Lopresto, C.G.; Calabro, V.; Curcio, S.; Chakraborty, S. Bioplastic from renewable biomass: A facile solution for a greener environment. *Earth Syst. Environ.* **2021**, *5*, 231–251. [CrossRef]
3. Wu, F.; Misra, M.; Mohanty, A.K. Challenges and new opportunities on barrier performance of biodegradable polymers for sustainable packaging. *Prog. Polym. Sci.* **2021**, *117*, 101395. [CrossRef]
4. Cabedo, L.; Feijoo, J.L.; Villanueva, M.P.; Lagar´on, J.M.; Gimenez, E. Optimization of biodegradable nanocomposites based on PLA/PCL blends for food packaging applications. *Macromol. Symp.* **2006**, *233*, 191–197. [CrossRef]
5. Yu, L.; Dean, K.; Li, L. Polymer blends and composites from renewable resources. *Prog. Polym. Sci.* **2006**, *31*, 576–602. [CrossRef]

6. Rhim, J.W.; Hong, S.I.; Ha, C.S. Tensile, water barrier and antimicrobial properties of PLA/nanoclay composite films. *LWT Food Sci. Technol.* **2009**, *42*, 612–617. [CrossRef]
7. Popa, I.; Offenberg, H.; Beldie, C.; Uglea, C.V. Benzocaine modified maleic anhydride copolymers-I. Synthesis and characterization of benzocaine modified poly(maleic anhydride-co-vinyl acetate), poly(maleic anhydride-co-methyl methacrylate) and poly(maleic anhydride-co-styrene). *Eur. Polym. J.* **1997**, *33*, 1511–1514. [CrossRef]
8. Wallach, J.A.; Huang, S.J. Copolymers of itaconic anhydride and methacrylate-terminated poly(lactic acid) macromonomers. *Biomacromolecules* **2000**, *1*, 174–179. [CrossRef]
9. Jeon, I.; Lee, S.W.; Jho, J.Y. Compatibilizing effect of poly(methyl methacrylate-co-maleic anhydride) on the morphology and mechanical properties of polyketone/polycarbonate blends. *Macromol. Res.* **2019**, *27*, 821–826. [CrossRef]
10. Bunkerd, R.; Molloy, R.; Punyodom, W.; Somsunan, R. Reactive blending of poly(L-lactide) and chemically-modified starch grafted with a maleic anhydride-methyl methacrylate copolymer. *Macromol. Symp.* **2015**, *354*, 340–346. [CrossRef]
11. Xu, J.; Shi, W.; Ming, G.; Fei, Y.; Yu, F. Preparation of poly(methyl methacrylate-co-maleic anhydride)/SiO_2-TiO_2 hybrid materials and their thermo- and photodegradation behaviors. *J. Appl. Polym. Sci.* **2005**, *97*, 1714–1724. [CrossRef]
12. Becker, D.; Hage, E.J.; Pessan, L.A. Synthesis and characterization of poly(methyl methacrylate-co-maleic anhydride) copolymers and its potential as a compatibilizer for amorphous polyamide blends. *J. Appl. Polym. Sci.* **2007**, *106*, 3248–3252. [CrossRef]
13. Xiao, L.; Li, Z.; Dong, J.; Liu, L.; Lei, S.; Zhang, X.; Zhang, H.; Ao, Y.; Dong, A. Fabrication of poly(methyl methacrylate-co-maleic anhydride) copolymers and their kinetic analysis of the thermal degradation. *Colloid Polym. Sci.* **2015**, *293*, 2807–2813. [CrossRef]
14. Dakka, S.M. TG/DTA/MS of poly(methyl methacrylate), the effect of particle size. *J. Therm. Anal. Calorim.* **2003**, *74*, 729–734. [CrossRef]
15. Wang, X.; Wu, L.; Li, J. Synergistic flame retarded poly(methyl methacrylate) by nano-ZrO_2 and triphenylphosphate. *J. Therm. Anal. Calorim.* **2011**, *103*, 741–746. [CrossRef]
16. Pal, M.K.; Singh, B.; Guatam, J. Thermal stability and UV-shielding of polymethyl methacrylate and polystyrene modified with calcium carbonate nanoparticles. *J. Therm. Anal. Calorim.* **2012**, *107*, 85–96. [CrossRef]
17. Lomonaco, D.; Maia, F.J.N.; Mazzetto, S.E. Thermal evaluation of cashew nutshell liquid as new bioadditives for poly(methyl methacrylate). *J. Therm. Anal. Calorim.* **2013**, *111*, 619–626. [CrossRef]
18. Cervantes-Uc, J.M.; Cauich-Rodriguez, J.V.; Herrera-Kao, W.A.; Vazquez-Torres, H.; Marcos-Fernandez, A. Thermal degradation behavior of polymethacrylates containing side groups. *Polym. Degrad. Stab.* **2008**, *93*, 1891–1900. [CrossRef]
19. Holland, B.J.; Hay, J.N. The effect of polymerization conditions on the kinetics and mechanisms of thermal degradation of PMMA. *Polym. Degrad. Stab.* **2002**, *77*, 435–439. [CrossRef]
20. Arshady, R. Suspension, emulsion, and dispersion polymerization: A methodological survey. *Colloid Polym. Sci.* **1992**, *270*, 717–732. [CrossRef]
21. Fitzwater, S.; Chang, H.R.; Parker, H.Y.; Westmoreland, D.G. Propagating radical termination at high conversion in emulsion polymerization of MMA: Rate coefficient determination from ESR data. *Macromolecules* **1999**, *32*, 3183–3189. [CrossRef]
22. Ishigami, A.; Watanabe, K.; Kurose, T.; Ito, H. Physical and morphological properties of tough and transparent PMMA-based blends modified with polyrotaxane. *Polymers* **2020**, *12*, 1790. [CrossRef]
23. Mohamed, D.; Fourati, Y.; Tarrés, Q.; Delgado-Aguilar, M.; Mutjé, P.; Boufi, S. Blends of PBAT with plasticized starch for packaging applications: Mechanical properties, rheological behaviour and biodegradability. *Ind. Crops Prod.* **2020**, *144*, 112061. [CrossRef]
24. Jiao, J.; Xiangbin, Z.; Xianbo, H. An overview on synthesis, properties and applications of poly(butylene-adipate-co-terephthalate)–PBAT. *Adv. Ind. Eng. Polym. Res.* **2020**, *3*, 19–26.
25. Kumar, S.; Krishnan, S.; Mohanty, S.; Nayak, S.K. PBAT-based blends and composites. In *Biodegradable Polymers, Blends and Composites*; Elsevier Inc.: Amsterdam, The Netherlands, 2022; pp. 327–354. [CrossRef]
26. Burford, T.; William, R.; Samy, M. Biodegradable poly(butylene adipate-co-terephthalate) (PBAT). In *Physical Sciences Reviews*; CRC Press: Boca Raton, FL, USA, 2021. [CrossRef]
27. Zhang, M.; Diao, X.; Jin, Y.; Weng, Y. Preparation and characterization of biodegradable blends of poly(3-hydroxybutyrate-co-3-hydroxyhexanoate) and poly(butylene adipate-co-terephthalate). *J. Polym. Eng.* **2016**, *36*, 473–480. [CrossRef]
28. Sargsyan, A.; Tonoyan, A.; Davtyan, S.; Schick, C. The amount of immobilized polymer in PMMA/SiO_2 nanocomposites determined from calorimetric data. *Eur. Polym. J.* **2007**, *43*, 3113–3127. [CrossRef]
29. Priestley, R.D.; Rittigstein, P.; Broadbelt, L.J.; Fukao, K.; Torkelson, J.M. Evidence for the molecular-scale origin of the suppression of physical ageing in confined polymer: Fluorescence and dielectric spectroscopy studies of polymer–silica nanocomposites. *J. Phys. Condens. Matter* **2007**, *19*, 205120–205132. [CrossRef]
30. Chrissafs, K.; Paraskevopoulos, K.M.; Pavlidou, E.; Bikiaris, D. Thermal degradation mechanism of HDPE nanocomposites containing fumed silica nanoparticles. *Thermochim. Acta* **2009**, *485*, 65–71. [CrossRef]
31. Voronin, E.F.; Gun'ko, V.M.; Guzenko, N.V.; Pakhlov, E.M.; Nosach, L.V.; Leboda, R.; Skubiszewska-Zieba, J.; Malysheva, M.L.; Borysenko, M.V.; Chuiko, A.A. Interaction of poly(ethylene oxide) with fumed silica. *J. Colloid Interface Sci.* **2004**, *279*, 326–340. [CrossRef]
32. Chrissafs, K.; Paraskevopoulos, K.M.; Papageorgiou, G.Z.; Bikiaris, D.N. Thermal and dynamic mechanical behavior of bio-nanocomposites: Fumed silica nanoparticles dispersed in poly(vinyl pyrrolidone), chitosan, and poly(vinyl alcohol). *J. Appl. Polym. Sci.* **2008**, *110*, 1739–1749. [CrossRef]

33. Lee, J.; Lee, K.J.; Jang, J. Effect of silica nanofillers on isothermal crystallization of poly(vinyl alcohol): In-situ ATR-FTIR study. *Polym. Test.* **2008**, *27*, 360–367. [CrossRef]
34. Chung, Y.-L.; Ansari, S.; Estevez, L.; Hayrapetyan, S.; Giannelis, E.P.; Lai, H.M. Preparation and properties of biodegradable starch-clay nanocomposites. *Carbohydr. Polym.* **2010**, *79*, 391–396. [CrossRef]
35. Glenn, G.; Klamczynski, A.; Ludvik, C.; Chiou, B.S.; Shed, I.; Shey, U.; William, O.; Delilah, W. In situ lamination of starch-based baked foam packaging with degradable films. *Packag. Technol. Sci.* **2007**, *20*, 77–85. [CrossRef]
36. Paul, D.R.; Robeson, L.M. Polymer nanotechnology: Nanocomposites. *Polymer* **2008**, *49*, 3187–3204. [CrossRef]
37. Venkatesan, R.; Rajeswari, N. ZnO/PBAT nanocomposite films: Investigation on the mechanical and biological activity for food packaging. *Polym. Adv. Technol.* **2017**, *28*, 20–27. [CrossRef]
38. Muthuraj, R.; Manjusri, M.; Mohanty, A.K. Biodegradable poly(butylene succinate) and poly(butylene adipate-co-terephthalate) blends: Reactive extrusion and performance evaluation. *J. Polym. Environ.* **2014**, *22*, 336–349. [CrossRef]
39. Venkatesan, R.; Rajeswari, N. Poly(butylene adipate-co-terephthalate) bionanocomposites: Effect of SnO_2 NPs on mechanical, thermal, morphological, and antimicrobial activity. *Adv. Compos. Hybrid Mater.* **2018**, *1*, 731–740. [CrossRef]
40. Schyns, Z.O.G.; Shaver, M.P. Mechanical recycling of packaging plastics: A review. *Macromol. Rapid Commun.* **2021**, *42*, e2000415. [CrossRef]
41. Fan, B.; Zhao, X.; Liu, Z.; Xiang, Y.; Zheng, X. Inter-component synergetic corrosion inhibition mechanism of *Passiflora edulia Sims* shell extract for mild steel in pickling solution: Experimental, DFT and reactive dynamics investigations. *Sustain. Chem. Pharm.* **2022**, *29*, 100821. [CrossRef]
42. Rabiee, H.; Ge, L.; Zhang, X.; Hu, S.; Li, M.; Yuan, Z. Gas diffusion electrodes (GDEs) for electrochemical reduction of carbon dioxide, carbon monoxide, and dinitrogen to value-added products: A review. *Energy Environ. Sci.* **2021**, *14*, 1959–2008. [CrossRef]
43. Zhou, S.; Zhai, X.; Zhang, R.; Wang, W.; Lim, L.-T.; Hou, H. High-throughput fabrication of antibacterial Starch/PBAT/AgNPs@SiO_2 films for food packaging. *Nanomaterials* **2021**, *11*, 3062. [CrossRef]
44. Thiyagu, T.T.; Gokilakrishnan, G.; Uvaraja, V.C.; Maridurai, T.; Arun Prakash, V.R. Effect of SiO_2/TiO_2 and ZnO nanoparticle on cardanol oil compatibilized PLA/PBAT biocomposite packaging film. *Silicon* **2022**, *14*, 3795–3808. [CrossRef]
45. Dubey, R.S.; Rajesh, Y.B.R.D.; More, M.A. Synthesis and characterization of SiO_2 nanoparticles via sol-gel method 725 for industrial applications. *Mater. Today Proc.* **2015**, *2*, 3575–3579.
46. Venkatesan, R.; Rajeswari, N. Preparation, mechanical and antimicrobial properties of SiO_2/poly(butylene adipate-co-terephthalate) films for active food packaging. *Silicon* **2019**, *11*, 2233–2239. [CrossRef]
47. Nasirtabrizi, M.H.; Ziaei, Z.M.; Jadid, A.P.; Fatin, L.Z. Synthesis and chemical modification of maleic anhydride copolymers with phthalimide groups. *Int. J. Ind. Chem.* **2013**, *4*, 11. [CrossRef]
48. Venkatesan, R.; Zhang, Y.; Che, G. Preparation of poly(butylene adipate-co-terephthalate)/$ZnSnO_3$ composites with enhanced antimicrobial activity. *Compos. Commun.* **2020**, *22*, 100469. [CrossRef]
49. Majoul, N.; Aouida, S.; Bessaïs, B. Progress of porous silicon APTES-functionalization by FTIR investigations. *Appl. Surf. Sci.* **2015**, *331*, 388–391. [CrossRef]
50. Rahman, I.A.; Padavettan, V. Synthesis of silica nanoparticles by sol-gel: Size-dependent properties, surface modification, and applications in silica-polymer nanocomposites—A review. *J. Nanomater.* **2012**, *2012*, 132424. [CrossRef]
51. Gao, Z.; Dong, X.; Li, L.; Ren, J. Novel two-dimensional silicon dioxide with in-plane negative poisson's ratio. *Nano Lett.* **2017**, *17*, 772–777. [CrossRef]
52. Kim, T.G.; An, G.S.; Han, J.S.; Hur, J.U.; Park, B.G.; Choi, S.C. Synthesis of size controlled spherical silica nanoparticles via sol-gel process within hydrophilic solvent. *J. Korean Ceram. Soc.* **2017**, *54*, 49–54. [CrossRef]
53. Zhan, P.; Chen, J.; Zheng, A.; Huang, T.; Shi, H.; Wei, D.; Xu, X.; Guan, Y. Preparation of methyl methacrylate-maleic anhydride copolymers via reactive extrusion by regulating the trommsdorff effect. *Mater. Res. Express.* **2019**, *6*, 025315. [CrossRef]
54. Venkatesan, R.; Rajeswari, N. Nanosilica-reinforced poly(butylene adipate-co-terephthalate) nanocomposites: Preparation, characterization and properties. *Polym. Bull.* **2019**, *76*, 4785–4801. [CrossRef]
55. Moustsfa, H.; El Kissi, N.; Abou-Kandil, A.I.; Abdel-Aziz, M.S.; Dufresne, A. PLA/PBAT bionanocomposites with antimicrobial natural rosin for green packaging. *ACS Appl. Mater. Interfaces* **2017**, *9*, 20132–20141. [CrossRef] [PubMed]
56. Zehetmeyer, G.; Meira, S.M.M.; Scheibel, J.M.; de Oliveira RV, B.; Brandelli, A.; Soares, R.M.D. Influence of melt processing on biodegradable nisin-PBAT films intended for active food packaging applications. *J. Appl. Polym. Sci.* **2015**, *133*, 43212. [CrossRef]
57. Venkatesan, R.; Surya, S.; Raorane, C.J.; Raj, V.; Kim, S.-C. Hydrophilic composites of chitosan with almond gum: Characterization and mechanical, and antimicrobial activity for compostable food packaging. *Antibiotics* **2022**, *11*, 1502. [CrossRef] [PubMed]
58. Jaramillo, A.F.; Riquelme, S.; Montoya, L.F.; Sánchez-Sanhueza, G.; Medinam, C.; Rojas, D.; Salazar, F.; Sanhueza, J.P.; Meléndrez, M.F. Influence of the concentration of copper nanoparticles on the thermo-mechanical and antibacterial properties of nanocomposites based on poly(butylene adipate-co-terephthalate). *Polym. Compos.* **2019**, *40*, 1870–1882. [CrossRef]

59. Jaramillo, A.F.; Riquelme, S.A.; Sánchez-Sanhueza, G.; Medina, C.; Solís-Pomar, F.; Rojas, D.; Montalba, C.; Melendrez, M.F.; Pérez-Tijerina, E. Comparative study of the antimicrobial effect of nanocomposites and composite based on poly(butylene adipate-*co*-terephthalate) using Cu and Cu/Cu$_2$O nanoparticles and CuSO$_4$. *Nanoscale Res. Lett.* **2019**, *14*, 158. [CrossRef]
60. Venkatesan, R.; Vanaraj, R.; Alagumalai, K.; Asrafali, S.P.; Raorane, C.J.; Raj, V.; Kim, S.-C. Thermoplastic starch composites reinforced with functionalized POSS: Fabrication, characterization, and evolution of mechanical, thermal and biological activities. *Antibiotics* **2022**, *11*, 1425. [CrossRef]

Disclaimer/Publisher's Note: The statements, opinions and data contained in all publications are solely those of the individual author(s) and contributor(s) and not of MDPI and/or the editor(s). MDPI and/or the editor(s) disclaim responsibility for any injury to people or property resulting from any ideas, methods, instructions or products referred to in the content.

Article

Evaluation of PVC-Type Insulation Foam Material for Cryogenic Applications

Dae-Hee Kim [1], Jeong-Hyeon Kim [2], Hee-Tae Kim [1], Jeong-Dae Kim [1], Cengizhan Uluduz [3], Minjung Kim [2], Seul-Kee Kim [2,*] and Jae-Myung Lee [1,2,*]

[1] Department of Naval Architecture and Ocean Engineering, Pusan National University, Busan 46241, Republic of Korea
[2] Hydrogen Ship Technology Center, Pusan National University, Busan 46241, Republic of Korea
[3] Diab Korea, A505 SKY-Biz Tower, 97 Centum Jungang-Ro, Haeundae-gu, Busan 48058, Republic of Korea
* Correspondence: skkim@pusan.ac.kr (S.-K.K.); jaemlee@pusan.ac.kr (J.-M.L.)

Abstract: With the International Maritime Organization (IMO) reinforcing environmental regulations on the shipbuilding industry, the demand for fuels, such as liquefied natural gas (LNG) and liquefied petroleum gas (LPG), has soared. Therefore, the demand for a Liquefied Gas Carrier for such LNG and LPG also increases. Recently, CCS carrier volume has been increasing, and damage to the lower CCS panel has occurred. To withstand liquefied gas loads, the CCSs should be fabricated using a material with improved mechanical strength and thermal performance compared with the conventional material. This study proposes a polyvinyl chloride (PVC)-type foam as an alternative to commercial polyurethane foam (PUF). The former material functions as both insulation and a support structure primarily for the LNG-carrier CCS. To investigate the effectiveness of the PVC-type foam for a low-temperature liquefied gas storage system, various cryogenic tests, namely tensile, compressive, impact, and thermal conductivity, are conducted. The results illustrate that the PVC-type foam proves stronger than PUF in mechanical performance (compressive, impact) across all temperatures. In the tensile test, there are reductions in strength with PVC-type foam but it meets CCS requirements. Therefore, it can serve as insulation and improve the overall CCS mechanical strength against increased loads under cryogenic temperatures. Additionally, PVC-type foam can serve as an alternative to other materials in various cryogenic applications.

Keywords: cargo containment system (CCS); mechanical characteristics; thermal conductivity; structure support; polyvinyl chloride (PVC)-type foam; polyurethane foam (PUF); cryogenic

1. Introduction

Since the MARPOL convention of 1973/1978, the International Maritime Organization (IMO) has implemented emission standards on the major environmental pollutants originating from vessels, namely NOx and SOx [1]. Recently, countries have been reinforcing these regulations to reduce the oxide content of ship fuels under the supervision of the IMO [2]. The increasingly restrictive regulations have spurred the development of several methods to reduce pollution, e.g., exhaust gas recirculation and selective catalytic reduction systems. However, the equipment cost of these methods is quite high. Hence, alternative forms of conventional fuels, which meet the abovementioned regulations, are being considered [3].

Natural gas (NG) is one such alternative. Generally, it can be used as liquefied natural gas (LNG). When NG is converted to LNG, its volume is reduced by more than 600-times. It is a clear, odorless, non-toxic, and non-corrosive cryogenic liquid at atmospheric pressure. Liquefied petroleum gas (LPG) is also a clean, abundant, and ecofriendly fuel. It can be easily condensed, packaged, stored, and utilized, and, hence, large amounts of energy can be stored and transported compactly. Therefore, the demand for LPG and LNG has increased, leading to a surge in the demand for LNG and LPG carriers [4,5]. The transportation and storage of LNG and LPG are significant concerns. LNG is transported at a cryogenic temperature (−163 °C), and LPG is transported at a subzero temperature (−42 °C). Additionally, they are stored in the state that they are transported in. Therefore, LNG and LPG carriers ought to be equipped with specially designed cargo containment systems (CCSs) to maintain cryogenic or subzero temperatures.

The LNG CCS is a membrane-type tank and is largely subdivided into two types: Mark III and NO96 [6]. As illustrated in Figure 1a, Mark III is composed of a primary stainless-steel 304 L membrane positioned on top of an insulation panel that incorporates a composite secondary barrier. The insulation consists of a reinforced polyurethane foam (R-PUF) [7]. In the NO96 system, both the primary and secondary barriers are made of thin sheets of invar, a Ni alloy. The shape of the box made of plywood filled with pearlite glass wool is shown in Figure 1b. As the NO96 panel boxes are contained in a sliding structure, the loads generated by the movement of the panels must be supported; this is achieved with the PVC-type foam located at the sidewall (Figure 1b). It simultaneously serves as an insulation and a liner to safely transport LNG at −163 °C [8].

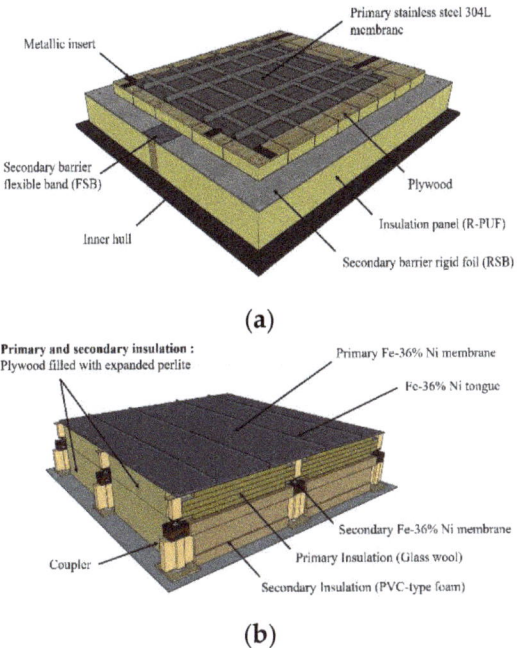

Figure 1. (a) Mark III and (b) NO96 liquefied natural gas (LNG) cargo containment system (CCS).

The LPG CCS is similar to the LNG CCS; however, the latter is an independent tank type A defined by the IMO. The cargo tank consists of outer shell plates, bulkheads, stiffeners, web frames, and stringers. The outside of the tank is insulated in a layer of polyurethane foam (PUF). Therefore, the insulation panel is a critical part of both LNG and LPG CCSs [9]. The inner panels of the LNG and LPG tanks have a large temperature difference from the outer ones and are subject to heavy loads. Proper insulating materials must be used to maintain the inner temperature and load resistance. A schematic of the LNG and LPG tanks is depicted in Figure 2 [10,11].

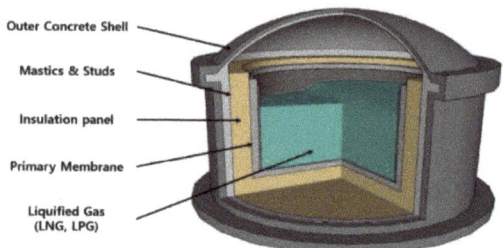

Figure 2. Approximate schematic diagram of a common storage tank (LNG and liquefied petroleum gas (LPG)).

As observed in the figure, all materials and structures in the primary membrane, insulation panel, etc., are the same as LNG CCS Mark III, except for R-PUF. PUFs are used as insulation for land storage tanks instead of R-PUF [11]. PUF and PVC foams are used as insulation materials in the CCS and storage tank. Currently, PUF is preferred as the main structural material in various cryogenic devices and has several applications—as insulation in cryogenic storage tanks for ship and rocket propulsion systems, core material in construction panels, and thermal insulation of refrigerated vehicles. Recently, South Korea has been developing an LNG CCS line called KC series, with PUF as an insulation material. PUF has already been employed for insulation in KC-1 [12].

As mentioned above, PUF finds applications in several fields. In particular, it is widely used in CCS. However, among them, engineers have observed stress concentration at the bottom of the insulation of the Mark III CCS. The insulation panels were attached to the inner walls of the containment using a series of mastic ropes (Figure 3) [13,14]. This phenomenon occurred because of a temperature gradient along the depth or height of the CCS as well as gaps between the mastic ropes (Figure 3) [15,16]. This phenomenon could cause severe damage or collapse of the structure. Therefore, in the lower part of the insulation panel, indicated by the red section in Figure 3, the mechanical strength of the material plays a more important role than the thermal performance. Larger carrier vessels and other larger ships are being commissioned to efficiently transport more consignments. With the increase in demand for LNG and LPG or heavier freight, the acting loads on the CCS also increase. Hence, the components of a CCS or storage tank require greater mechanical strength and thermal performance [17]. Therefore, evaluation of the strength of the insulation panel structures is important [18].

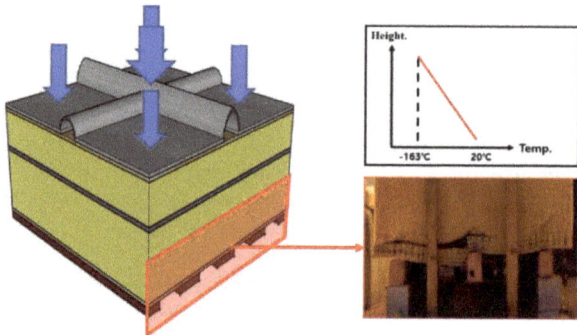

Figure 3. Stress concentration at the bottom of an insulation panel and the corresponding temperature gradient.

Accordingly, PVC-type foam is used to improve the mechanical strength of NO96-type LNG CCS as it is a high-strength, low-weight polymeric material, with closed cell structures. Hence, it has been widely used as an insulation and reinforcing material for LNG CCS.

In the past few decades, several studies have been conducted to evaluate the mechanical strength of polymer insulation materials under cryogenic environments. Park et al. (2016) investigated the effect of the blowing agent on the cryogenic temperature mechanical properties of glass-fiber-reinforced polyurethane foams. The test was conducted at various cryogenic temperatures and strain rates. The compressive elastic modulus and compressive strength increased as the temperature decreased. At $-163\ °C$, the fluctuations became more severe than those at other temperatures. In the aspect of blowing agent, the compressive stress of the CO2-blown reinforced polymer foam was higher than that of the HFC-245fa-blown reinforced polyurethane foam [19]. Son et al. (2019) investigated the synthesis of polyurethane foam for application to cryogenic environments. Polyurethane foams were manufactured, and samples were subjected to mechanical tests to investigate the mechanical properties of the polyurethane foam within a temperature range of $25\ °C$ to $-163\ °C$. In addition, thermal and microstructural investigations were conducted. The compression behavior of polyurethane foam was also analyzed. Regardless of the blowing agent content, all the polyurethane foams showed better compression strength at cryogenic temperatures than at room temperature. The compression strength was improved from at least 76% to 199% [20]. Marsavina and Linul (2022) investigated the fracture toughness of rigid polymeric foams. The different approaches for estimating the fracture toughness of polymeric foams were reviewed. Focus was given on the parameters influencing the fracture toughness of polymeric foams, such as specimen type, solid material, density, loading speed, size effect, and temperature [21]. Including representative research activities, several research outcomes have been reported on the effect of cryogenic temperatures on the mechanical characteristics of polymeric foams [21]. However, most of the research has been focused on polyurethane foam, which is already applied in cryogenic liquid gas storage systems [22–28]. Few studies have reported the mechanical properties and performance of PVC-type foam under cryogenic environments [8].

The primary objective of this study is to closely examine the material properties and insulation performance of the PVC-type foam under cryogenic temperatures by conducting tensile, compressive, and impact tests. In addition, thermal conductivity, which is an important material property of the foam insulation, was investigated. Finally, the criteria for the further applications of the PVC-type foam in various fields are presented.

2. Materials and Methods

2.1. Test Specimen and Apparatus

To investigate the mechanical properties of the PVC-type foam, thermal conductivity, compressive, tensile, and impact tests were conducted under ambient and cryogenic temperatures. The PVC-type foam was commercially manufactured (Diab, Helsingborg, Sweden). The tested PVC foam material was manufactured using PVC, isocyanates, anhydrides, and blowing agents. The PVC resin is a thermoplastic component, and isocyanates contribute to the fabrication of a new thermoplastic polymer as well as a thermoset to improve mechanical properties. In addition, anhydride is a plasticizer and contributor to the gas phase. The PVC foam is manufactured in various steps. The raw materials are mixed to a homogeneous plastisol. Then, the plastisol is filled into molds, which are then put into presses and subjected to heat and pressure. This creates a rubbery, small, and visually solid block. Then, it is expanded and further cured in heated water and a steam environment in different chambers with varying temperatures. In addition, polyurethane foam was manufactured directly by synthesizing basic raw materials, such as polyol, isocyanate, and blowing agents. The polyol and isocyanate were mixed at a fixed ratio of 700:800. The polyol, isocyanate, and blowing agent were mixed by a homogenizer at 4000 rpm for approximately 60 s to synthesize a homogenous PUF solution. Then, the solution was poured into an open mold (length × width × height = 300 mm × 300 mm × 250 mm). It was then foamed and cured at ambient temperature (20 °C). Finally, after 24 h, the solid PUF was machined to a different size for each test [29,30]. The dimensions of the thermal conductivity test specimen were 280 mm × 280 mm × 25 mm, while those of the compressive and impact test specimen were 50 mm × 50 mm × 25 mm, according to KS M ISO 844 and KS M ISO 6603. The tensile test specimen was also manufactured according to KS M ISO 1926. Table 1 shows the average density of the PUF and PVC-type foam for each test. A universal testing machine (UTM) with a cryogenic chamber was used for the compressive and tensile tests. The reason for using the cryogenic chamber was to maintain the cryogenic conditions with liquid nitrogen (LN2) and an automatic temperature control box. The thermal conductivity was measured using a heat flow meter (HFM436, NETZSCH).

Table 1. Average density of the PUF and PVC-type foam for each test.

Test Method	Test Material	Density (kg/mm^3)	Specimen Thickness (mm)
Insulation	PUF	68.8 (L)	25
		138.4 (H)	
	PVC-type foam	61.1 (L)	
		138.1 (H)	
Compressive & Impact Test	PUF	64 (L)	25
		144 (H)	
	PVC-type foam	64 (L)	
		144 (H)	
Tensile test	PUF	64 (L)	15
		144 (H)	
	PVC-type foam	64 (L)	
		144 (H)	

2.1.1. Thermal Conductivity

The thermal conductivity was measured using a heat flow meter HFM436 (NETZSCH). This equipment consisted of upper and lower plates; the upper plate had a higher temperature than the lower plate, and the specimen was placed between the plates. The dimensions of the specimen were depicted before. Subsequently, heat flow was generated to measure the thermal conductivity according to KS L 9016.

2.1.2. Tensile Test

In the tensile test, the specimens were precooled according to the test temperature. Subsequently, they were positioned in the UTM equipped with a tensile jig. After attaching the strain gauge, the tensile test was performed at a tensile test speed of (5 ± 1) mm/min according to the KS M ISO 1926.

2.1.3. Compressive Test

In the compressive test, after the specimens were precooled, similar to the tensile test, they were positioned in the UTM equipped with a compressive jig. Subsequently, the test speed was set to 2.5 mm/min, which was approximately 10% of the thickness of the specimen according to KS M ISO 844, and the experiment was conducted.

2.1.4. Impact Test

The impact test was conducted on Instron CEAST9340. The test was conducted using two different impact energies fitted to two specimens with different densities under a varying temperature, ranging from room to cryogenic temperatures. The criterion for selecting the impact energy used energy absorption as a measure for estimating the impact performance. To calculate the absorption energy, the force before the densification region was required [31]. Therefore, the two impact energies near the densification region were selected in this study and determined to be 30 and 60 J.

Subsequently, a square impactor, which had a cross-section similar to that of the specimen, was used. The weight mounted on the impact equipment was approximately 30 kg. The low-temperature impact test specimens were precooled for approximately 1 h before the impact experiment, similar to the tensile and compressive tests. The impact experiments were conducted within 5 s of removing the specimens from the cooled area, in accordance with KS M ISO 6603.

2.1.5. Scanning Electron Microscopy

Foam density is a very important parameter because it alters the number and size of the cell. Hence, the present study controlled the foam density by controlling the chemical reaction between blowing agent and isocyanate [32–34]. The cell morphology of the PUFs and PVC-type foams was observed using scanning electron microscopy (SEM; SUPRA25 (ZEISS, Jena, Germany) coupled with an energy-dispersive X-ray spectrometer). The SEM consisted of an electron gun, lens, chamber, and detector. The electron gun generated an electron beam and focused it on the inspection object at 10 kV. In addition, the lens controlled the size and intensity of the beam. Testing samples for SEM analysis were cut to 10 mm × 10 mm × 5 mm by a sharp razor blade to minimize damage to the cell structures [35,36]. The prepared testing samples were coated for protection against heated electron beam.

2.2. Experiment Temperature Environment

This study calculates the insulation performance and mechanical properties of the PVC-type foam as compared to PUF by varying the temperature and density. The temperature was selected based on the liquefaction temperature of LNG (−163 °C) and LPG (−42 °C). Therefore, the test temperatures were 20, −50, −110, and −170 °C. Particularly, in a cryogenic experiment, to calculate the exact mechanical properties, the internal temperatures of the test specimen should be equal to the external temperature. Therefore, specimens were precooled to each temperature prior to testing.

3. Results and Discussion
3.1. Cellular Morphology of PUF and PVC-Type Foam

The changes in mechanical properties mentioned above are a result of the changes in the cell structure. Hence, the cell morphology had to be verified. Figure 4 displays the SEM images of the PUF (L, H) and PVC-type foam (L, H) at approximately 32× magnification. Photographs of the cell sizes and shapes of the specimens are presented in Figure 4.

The cell size was different for specimens with different densities. It should be, and sometimes is, an important parameter. However, most mechanical and thermal properties have a weak dependence on the cell size, and the cell shape is more important [34]. On the one hand, the PVC-type foam and the PUF specimens have a closed cell shape, according to previous studies [37], and the cell shape of PVC-type foam is polyhedral in topology, like the typical PVC foam. On the other hand, the cell shape of the PUF was approximately spherical and bears the same appearance as the typical polyurethane foam [37–41]

Table 2 summarizes the cell diameters of PVC-type foams and PUFs. Previous studies confirmed that if the PUF (rigid PUF) had larger diameters, it also had higher thermal conductivity [42,43]. PUF (L) had a larger diameter than PUF (H), which is consistent with previous studies. However, PVC-type foam exhibited the opposite result to the PUF. PVC-type foam (L) had a smaller diameter than PVC-type foam (H). At low densities, the PUF (L) and PVC-type foam (L) had nearly identical diameters. However, at high densities, PVC-type foam (H) had approximately 120% larger diameter than neat PUFs. Hence, PVC-type foam and PUF have different cell size tendencies.

Table 2. Cell diameter sizes of neat PUFs and PVC-type foams.

Material Type	Density (kg/m^3)	Avg. Diameter (μm)	Standard Deviation	Max. Diameter (μm)
PUF	144 (H)	227.660	55.182	306.667
	64 (L)	388.138	92.211	581.513
PVC-type foam	144 (H)	509.957	118.880	522.595
	64 (L)	383.646	67.543	306.667

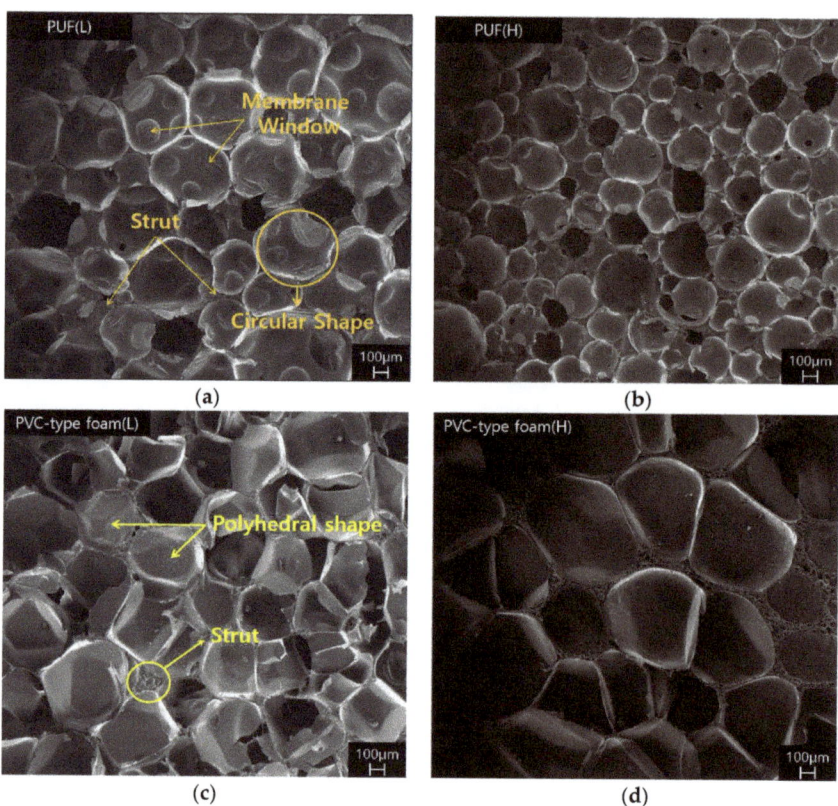

Figure 4. SEM images of microstructures: (a,b) PUF (L, H) and (c,d) PVC-type foam (L, H).

Figure 5 presents the SEM images of PVC-type foam observed at approximately 82× magnification. As presented in Figure 5, small cells of different sizes are generated within a small space between the formed cells. These shapes can be observed when the cells are random and of different sizes. Cells coarsen if the cells become connected (pore wall between them breaks) or if gas is generated in the cell. Subsequently, the many-sided cells grow, while the few-sided ones shrink, and this consequently results in an increasingly inhomogeneous structure. In this process, the diffusion rate of the cells is affected by pressure [34]. Therefore, unlike the PUF that was foamed and cured at ambient temperatures, PVC-type foam was processed at high pressures and temperatures. Therefore, while the cells were foaming, the cell diffusion rate was affected by the processing heat and pressure. Hence, this morphology appeared in PVC-type foam. These shapes can also be observed in Figure 6 at approximately 82× magnification. A few cell walls or cells in the PUF and PVC-type foam were damaged. The yellow circles in Figure 6 are the parts of the sparse distribution of cell sizes from Figure 5 that were confirmed to not have suffered severe damage and that retained the original shape. Therefore, the fact that these parts affected the compressive and impact strengths under a varying temperature (from room to cryogenic) is verified.

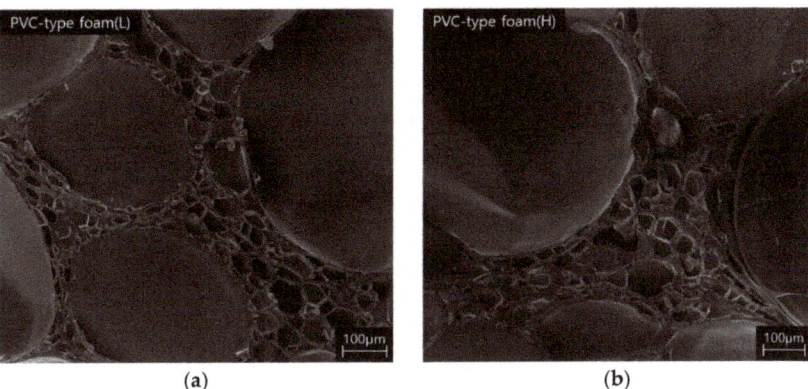

Figure 5. Micrograph of a PVC-type foam with a wide distribution of cell sizes: (**a**) PVC-type foam (L) and (**b**) PVC-type foam (H).

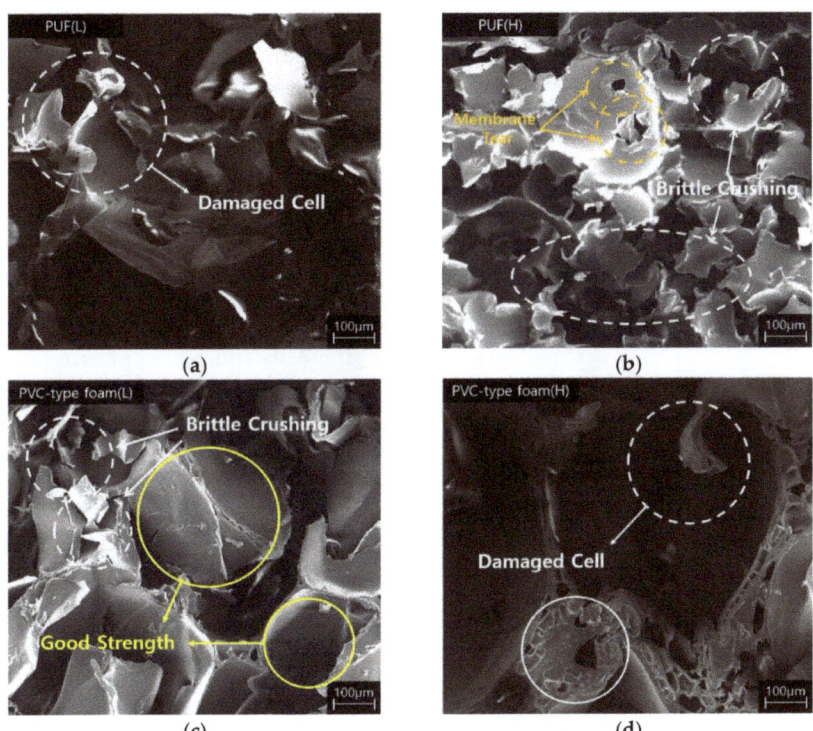

Figure 6. Micrographs of the specimens after the cryogenic compressive test (−170 °C): (**a**) PUF (L), (**b**) PUF (H), (**c**) PVC-type foam (L), and (**d**) PVC-type foam (H).

As mentioned in Section 2, the production of the foam-type material involves a blowing process. The thermal and mechanical properties may vary with the amounts of the true material and blowing agent. This is called the true density, which allows one to know the amount of the actual material. The results of the true density calculations of this PVC-type foam and PUF are listed in Table 3.

Table 3. Results of the true densities of the two types of materials.

Material Type	Average True Volume (cm^3)	Standard Deviation	Average True Density (kg/m^3)	Standard Deviation
PUF (H)	46.6257	0.2158	0.1480	0.0007
PVC-type foam (H)	46.3865	0.0059	0.1492	0.0000
PUF (L)	50.1830	0.1492	0.0729	0.0002
PVC-type foam (L)	54.1881	0.0245	0.0650	0.0000

At low densities, the PVC-type foam had a larger true volume occupied by the material relative to the similar mass than that of the PUF. Hence, its true density is lower than that of neat PUF. In the case of a high density, both exhibit similar values. As mentioned above, as the SEM images and cell size had different characteristics according to the densities of the two materials, the true density will also differ. Moreover, we considered that the true density affects the strength of the cryogenic environment for materials with high densities.

3.2. Thermal Conductivity

In general, thermal conductivity determines the insulating performance of the material. Materials with a high thermal conductivity are used as heat dissipaters, whereas materials with a low thermal conductivity are used as insulation. Therefore, we measured the thermal conductivity of the PVC-type foam and PUF using the HFM. In general, density is one of the most important parameters for determining the mechanical and thermal properties of foam materials. The low density of foam materials results in low thermal conductivity, thereby providing better insulating performance. To compare the thermal conductivity of PUF and PVC-type foam, foam materials with similar densities were selected. Figure 7 shows a comparison of thermal conductivity between PVC foam and neat PUF. In addition, Table 4 presents the measured thermal conductivity of PVC-type foam and neat PUF. Results showed that the thermal conductivity of neat PUF was lower than that of PVC-type foam. As shown in Figure 7, the difference in thermal conductivity between PUF and PVC-type foam is much more significant with increasing density. At low densities, the thermal conductivity of PUF was approximately 20% lower than that of PVC-type foam. In addition, the thermal conductivity of neat PUF was approximately 35% lower than that of PVC-type foam at relatively high densities.

Table 4. Thermal conductivity of PVC-type foam and neat PUF.

Material	Thermal Conductivity (Average)	Average Density (kg/mm^3)
PUF (H)	0.0327	138.4
PVC-type foam (H)	0.0447	138.1
PUF (L)	0.0262	68.8
PVC-type foam (L)	0.0320	62.1

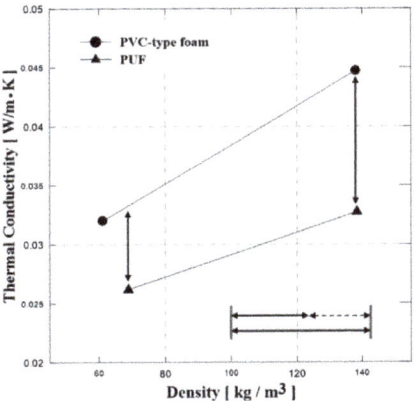

Figure 7. Comparison of thermal conductivity between PVC-type foam and PUF.

3.3. Tensile Test

Figure 8 shows the results of the test. Generally, polymer foam materials with higher densities possess higher strength than those with lower densities [44]. Therefore, as illustrated in Figure 8, the tensile strengths of both PVC-type foam and PUF, which had higher densities, were better than those with lower densities.

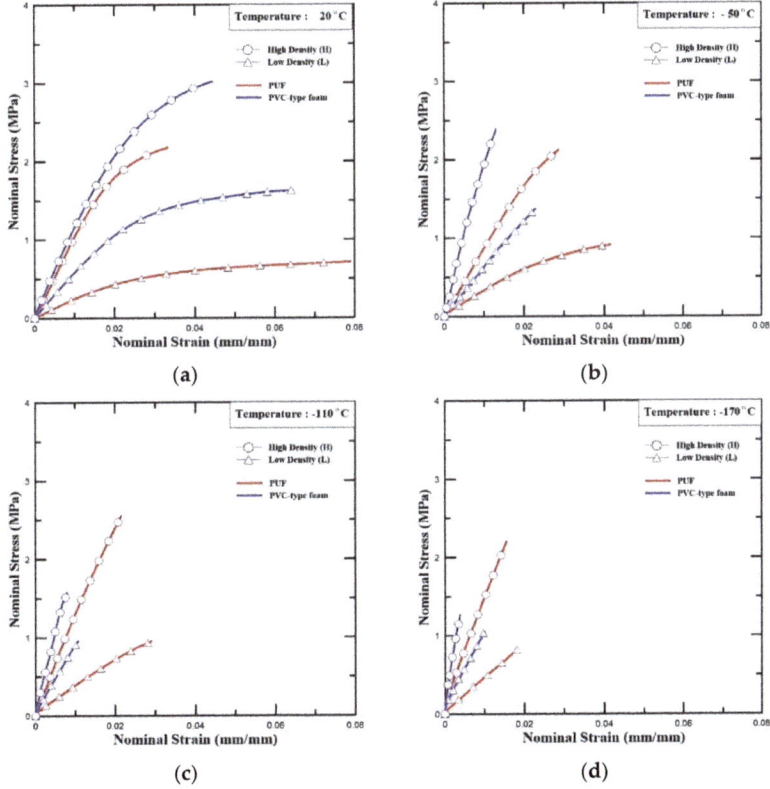

Figure 8. Stress–strain curve of the PUF (H, L) and PVC-type foam (H, L) with different temperatures: (**a**) 20 °C, (**b**) −50 °C, (**c**) −110 °C, and (**d**) −170 °C (tensile test).

In Figure 8, at room temperature and −50 °C, the tensile strength of PVC-type foam was confirmed to be comparatively higher than that of PUF. However, in contrast, the tensile strength of PUF (H) was higher than that of PVC-type foam (H) at low temperatures (−110 °C and −170 °C).

Figure 9 illustrates the trend of the PVC-type foam, indicated by blue lines, which decrease steadily. The curve decreased between −50 and −110 °C. The red lines of the PUF increase until −110 °C and then decrease. The yellow grid line in Figure 9 represents the tensile strength requirement of the insulation material when using R-PUF insulation in the CCS [7]. Except for low-density PUF, PVC-type foam (H, L) and PUF (H) were located higher than the yellow grid line, indicating the requirements for CCS. Therefore, although the strength of the PVC-type foam decreased as the temperature decreased, it satisfied the standard strength criterion.

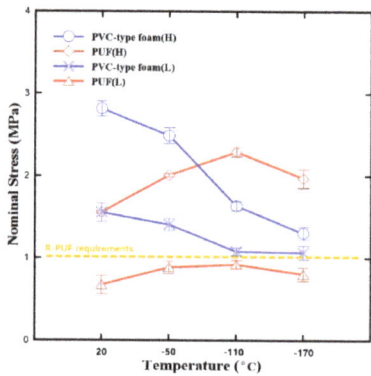

Figure 9. Nominal stress vs. temperature diagram comparing the tensile strengths of PVC-type foam and the PUF specimens with respect to R-PUF requirements at low and high temperatures.

Table 5 lists the tensile elastic modulus and yield stress of the specimens. The values in the table are based on the averaged test results. In the case of PUFs, the yield stress increased to −110 °C and had the highest yield stress at that temperature. In contrast, the yield stress of PVC-type foam tended to decrease as the temperature decreased.

Table 5. Test results of tensile Young's modulus and yield stress.

Material	Temperature	Young's Modulus (MPa)	Standard Deviation	Yield Stress (MPa)	Standard Deviation
PUF (L)	20 °C	19.347	1.883	0.672	0.074
	−50 °C	25.759	2.073	0.884	0.049
	−110 °C	33.376	2.671	0.919	0.033
	−170 °C	44.178	1.138	0.802	0.056
PUF (H)	20 °C	67.361	2.847	1.555	0.065
	−50 °C	89.619	4.563	2.007	0.097
	−110 °C	104.698	37.951	2.287	0.266
	−170 °C	142.142	17.354	1.968	0.563
PVC-type foam (L)	20 °C	50.939	0.814	1.555	0.065
	−50 °C	60.154	5.899	1.407	0.054
	−110 °C	79.404	5.880	1.076	0.186
	−170 °C	116.492	14.709	1.066	0.099
PVC-type foam (H)	20 °C	95.162	5.915	2.813	0.232
	−50 °C	170.611	12.569	2.487	0.241
	−110 °C	183.312	21.274	1.634	0.135
	−170 °C	306.282	16.078	1.302	0.198

As shown in Figure 8 and Table 5, the elastic modulus increases as the temperature decreases. The tensile elastic modulus increased up to 130% and 220% for the low densities and high densities, respectively, of the PVC-type foam at cryogenic temperatures. In the strength aspect, the lower the temperature, the weaker the PVC-type foams. Although it had lower strength at lower temperatures, its strength was higher than the neat PUF's highest strength at low density [45].

3.4. Compressive Test

Table 6 contains photographs of specimens after the compressive test at room and cryogenic temperatures, where the surfaces of specimens with the most marked differences are indicated. The specimens of both PUF and PVC-type foam were evenly deformed without any damage/failure at room to low temperatures (-50 to -110 °C). However, the breakage can be clearly observed at -170 °C in the cryogenic range. The PUF and PVC-type foam specimens were crushed and fractured. However, the neat PUF specimen was more severely crushed and fractured than the PVC-type foam. This phenomenon is due to the embrittlement of cellular structures at cryogenic temperatures [6–8]. Therefore, it can be confirmed that there is a critical point of cellular structure embrittlement, which is approximately -170°C. Compared with these results, the cellular structure is not severely embrittled in the PVC-type foam. Subsequently, a strain–stress curve with experimental data is plotted in Figure 10. In Figure 10, the left-hand-side axis is the nominal stress, which is expressed as follows:

$$\sigma = \frac{P}{A_0} \qquad (1)$$

where P is the load applied to the specimen and A_0 is the initial section area. The nominal compressive strain is expressed as follows:

$$\varepsilon = \frac{h_0 - h}{h_0} \qquad (2)$$

where h is the height of the specimen after undergoing a strain ε and h_0 is the original height. We observed three characteristic phases: a linear elastic region, plateau region, and region of sharp increase. The linear elastic region is used for calculating the elastic modulus. A small decrease called strain softening occurred before the plateau region. The stress strongly fluctuated and unexpectedly decreased—a phenomenon known as brittle crushing. These characteristics became more prominent as the experimental temperature was reduced owing to the embrittlement of the cell wall. The sharp rise occurred due to densification because the cells of the material collapsed and became compressed [41,45,46].

The four foam specimens exhibited better compressive performance at low temperatures than at room temperature, which is in line with the results of previous studies [46]. Therefore, in Figure 10, the compressive strengths and elastic modulus of PUF and PVC-type foam increase as the temperature decreases. Subsequently, as observed in Figure 10, PVC-type foam had a higher compressive strength than neat PUF. Further, the gap between PVC-type foam and PUF with a similar density tended to narrow as the temperature decreased. The linear slope at the beginning of the graph is called the elastic modulus, and the elastic modulus of the PVC-type foam is generally larger than the PUF at all temperatures. The average compressive strength was also greater than that of the PUF having the same density. These results are confirmed not only by Figure 10 but also by Table 7, highlighting the average values of the test results. Compressive yield stresses of the PUF were enhanced by approximately 150% at cryogenic temperatures than at room temperature. Likewise, the compressive elastic modulus also increased by approximately 100%. Yield stresses of the PVC-type foam increased by approximately 70% at cryogenic temperatures for low densities. Similarly, the compressive elastic modulus also increased by approximately 100%.

Table 6. Photographs of specimens after the compressive test.

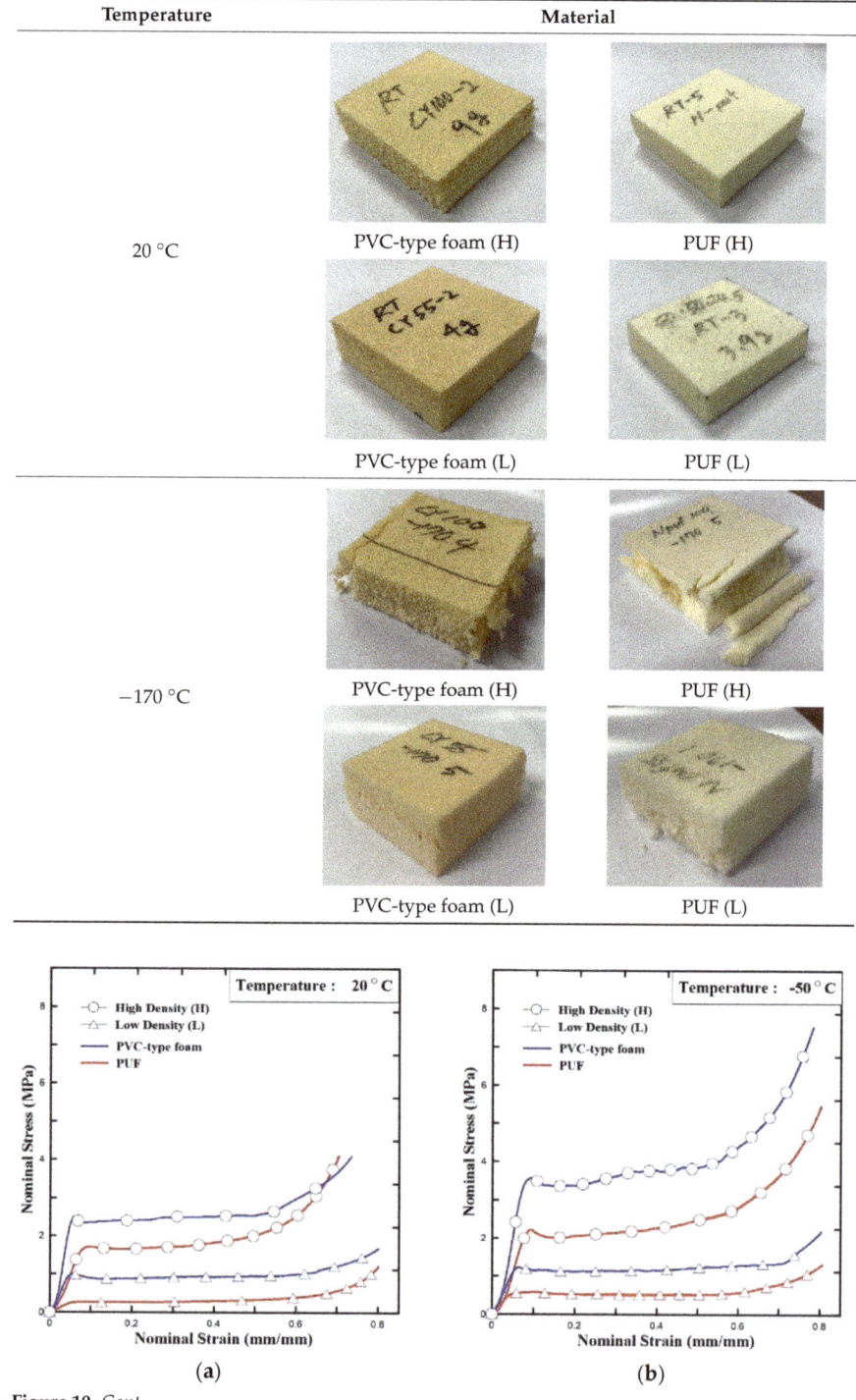

Figure 10. Cont.

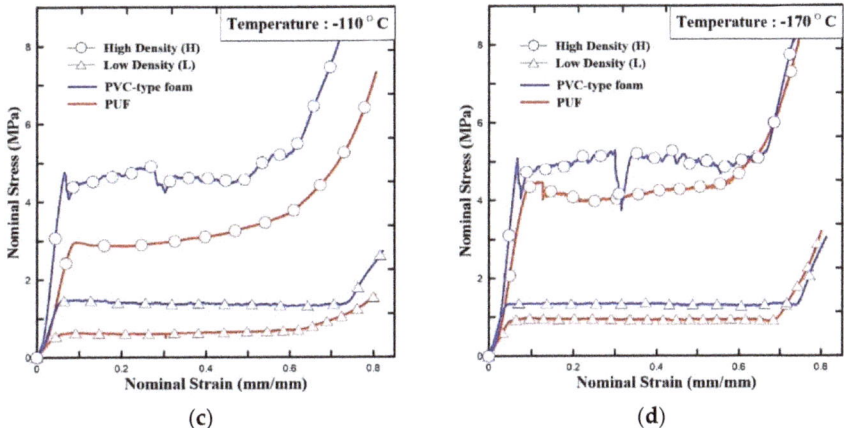

Figure 10. Stress–strain curve of the PUF(H, L) and PVC-type foam(H, L) at different temperatures: (**a**) 20 °C, (**b**) −50 °C, (**c**) −110 °C, and (**d**) −170 °C (compressive test).

Table 7. Test results of compressive Young's modulus and yield stress.

Material	Temperature	Young's Modulus (MPa)	Standard Deviation	Yield Stress (MPa)	Standard Deviation
PUF (L)	20 °C	6.537	2.729	0.337	0.060
	−50 °C	9.097	0.839	0.639	0.070
	−110 °C	9.857	0.079	0.722	0.079
	−170 °C	13.626	3.444	0.857	0.175
PUF (H)	20 °C	21.767	4.012	1.710	0.065
	−50 °C	23.644	2.103	2.188	0.121
	−110 °C	39.481	4.742	2.973	0.119
	−170 °C	43.557	2.524	4.342	0.271
PVC-type foam (L)	20 °C	17.259	4.306	0.870	0.284
	−50 °C	22.433	1.640	1.265	0.033
	−110 °C	24.947	2.106	1.400	0.051
	−170 °C	34.523	4.031	1.456	0.067
PVC-type foam (H)	20 °C	41.474	4.148	2.469	0.029
	−50 °C	55.169	6.119	3.600	0.063
	−110 °C	64.573	10.152	4.442	0.337
	−170 °C	73.322	2.083	4.738	0.285

Two standard compressive strength criteria exist for the optimal CCS using R-PUF at 20 °C and −170 °C [7]. The standard for −170 °C is higher than that for 20 °C as the embrittlement occurred. As indicated by the results in Figure 11, neither neat PUF nor PVC-type foam meet the criteria at low densities. However, as the generally used insulation density for the CCS is more similar to high density, the compressive stress of both neat PUF and PVC-type foam at high density satisfied both the two standard criteria for all temperatures.

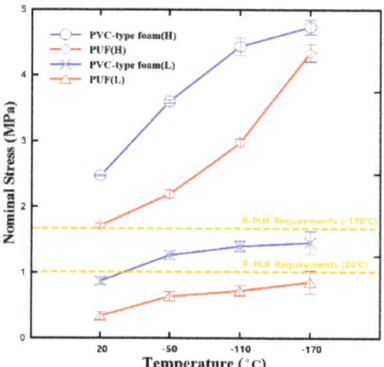

Figure 11. Nominal stress vs. temperature diagram comparing the compressive strengths of PVC-type foam and the PUF specimens with respect to R-PUF requirements at low and high temperatures.

3.5. Impact Behavior

Larger cargo ships have greater freight capacity. Therefore, the contents of the tank have a significant impact on not only the vessel's tank but also on the large storage tanks on land. This necessitates impact performance data, and, hence, the impact test was conducted. The results are presented in Figure 12.

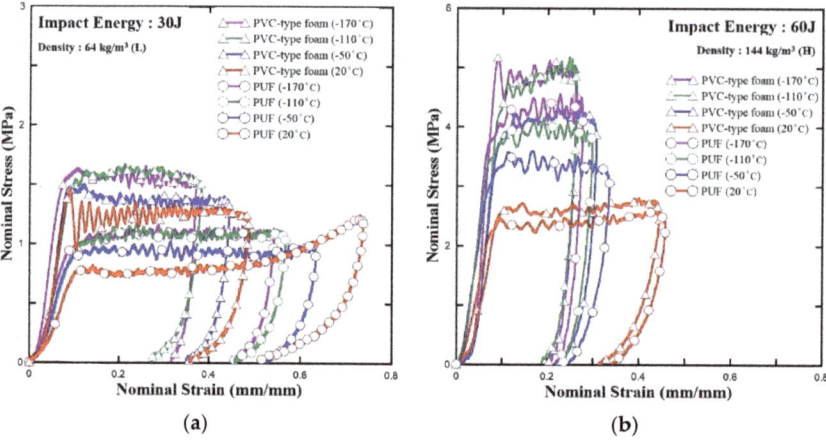

Figure 12. Stress–strain curve based on the results of the impact test: (**a**) 30 J and (**b**) 60 J of impact energy.

As presented in Figures 12 and 13 have four regions: linear, plateau, densification, and unloading. This figure shows the typical impact behavior of a foam-type material [47]. Similar to the abovementioned elastic region, it maintained linearity because the walls of the cells were bent and stretched [48]. Fluctuations occurred in the plateau region as the cells were damaged and crumbled by buckling. Subsequently, the cell walls and cell struts crumpled together. This caused the creation of the densification region. Next, the appearance of this region confirmed that stress rapidly decreased in the unloading region as the initial dynamic energy was removed [29]. Notably, this is similar to the compressive behavior. In Figure 12, the PVC-type foam is stronger than the PUF at both densities. Moreover, if the PUF and PVC-type foam have high density, they have higher strength. However, as the temperature increases, the strength decreases since the material becomes more brittle with a decrease in temperature. Therefore, the linear modulus of elasticity and yield stress increased.

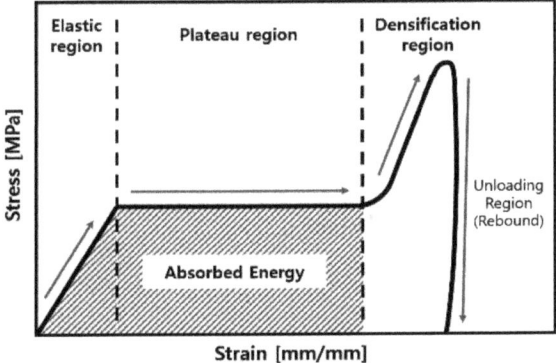

Figure 13. Stress–strain curve based on the results of the impact test.

Figure 13 illustrates the force–displacement curve of the PUFs and PVC-type foams with different densities at various temperatures and impact energies. Between the elastic region and the early segment of the plateau region, a point exists where the curve gradually linearly increases and then sharply decreases. This point is called the initial peak point, and the corresponding force is called the initial peak force. This force also increases as the density increases and the test temperature decreases [31].

Table 8 lists the average values of the test results. The higher the initial peak force, the higher the yield stresses and impact elastic modulus. Similar to previous studies, the findings of our study confirm that the PUFs and PVC-type foams both have a higher initial peak force if the temperature is reduced and the density is increased. Subsequently, PVC-type foams had a higher initial peak force than the PUFs under all the test conditions (Figure 12 and Table 8).

Table 8. Test results of initial peak force and ratio of absorbed energy.

Material	Temperature	Initial Peak Force (N)	Standard Deviation	Ratio of Absorbed Energy (%)	Standard Deviation
PUF (L)	20 °C	1924.701	126.129	88.218	0.130
	−50 °C	2091.565	77.305	93.306	0.434
	−110 °C	2408.992	304.272	94.451	1.050
	−170 °C	2738.886	9.966	95.307	0.329
PUF (H)	20 °C	6464.576	443.486	91.093	0.250
	−50 °C	7467.204	328.476	92.208	0.420
	−110 °C	9016.458	895.106	90.669	1.942
	−170 °C	9331.487	603.999	92.927	0.755
PVC-type foam (L)	20 °C	3633.627	36.618	93.106	0.048
	−50 °C	3749.665	66.690	93.889	0.623
	−110 °C	3906.940	426.257	93.335	1.608
	−170 °C	4079.559	66.358	94.803	0.191
PVC-type foam (H)	20 °C	6446.355	464.340	90.407	0.552
	−50 °C	10602.630	374.243	91.952	0.790
	−110 °C	11703.560	426.257	91.283	2.416
	−170 °C	12278.953	559.559	92.893	0.527

PVC-type foam also exhibited similar values with the PUF with high densities at room temperature. However, the values of PVC-type foam were evidently higher than that of the PUF by approximately 30% at −170 °C, while PVC-type foam in low density had a higher initial peak force up to approximately 88% at room temperature and up to approximately 48% at cryogenic temperature −170 °C. Although the differences were decreased by decreasing the test temperature, peak force of the PVC-type foams was higher than that of the PUF, as shown in Figure 14.

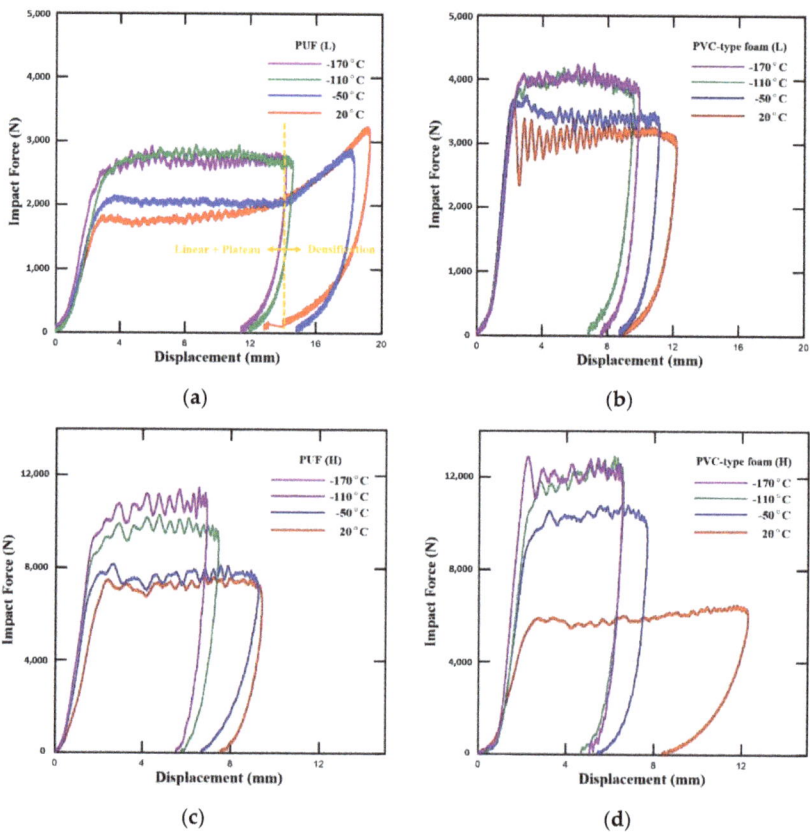

Figure 14. Force–displacement curves at 30 J (**a**,**b**) and 60 J (**c**,**d**) based on temperature: (**a**) neat PUF (L), (**b**) PVC-type foam (L), (**c**) Neat PUF (H), (**d**) PVC-type foam (H).

Table 8 and the left-hand side of Figure 15 present the absorption energy. The energy absorbed per unit volume due to a strain ε is calculated by the area under the stress–strain curve. It can be expressed as follows:

$$E_{absorbed} = \int_0^\varepsilon \sigma(\varepsilon)\, d\varepsilon, \tag{3}$$

where $\sigma(\varepsilon)$ is the stress as a function of strain, and the strain range is 0—ε, which is immediately before the start of the densification. When tested with the initial set impact energy, if the specimens are completely destroyed in the plateau region, the absorption energy can be obtained by integrating the stress value in the strain from the beginning to the end. This energy represents the area of the closed section of the graph. The obtained absorption energy is expressed as a ratio of the total impact energy. As the test temperature decreases, the ratio increases. This tendency was observed in previous studies as well, which tested composite or polymeric materials in cryogenic environments [48].

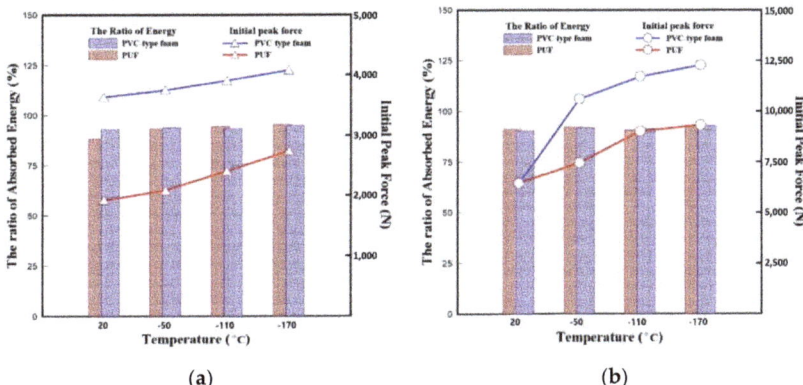

Figure 15. Comparison of specimens in terms of the initial peak force and the ratio of the absorbed energy by temperature: (**a**) for low densities (L), (**b**) for high densities (H).

4. Conclusions

This study investigated the mechanical and thermal characteristics of neat PUF and PVC-type foam, which offer different densities from room to cryogenic temperatures. The results obtained from the present study are summarized as follows.

- The thermal conductivity test indicates that PUF offers better insulation performance and presents a thermal conductivity approximately 35% lower at a density of 144 kg/m^3 and 20% lower at a density of approximately 64 kg/m^3 than PVC-type foam.
- The results of a compressive test on PVC-type foam with a high density (144 kg/m^3) exhibited that brittle crushing causes a significant decrease in target material strength. Additionally, the neat PUF tends to splinter into several pieces, whereas PVC-type foam develops dents. Thus, the PVC-type foams possess elevated strength over the PUFs.
- In the tensile test, the tensile strength and elastic modulus increase as temperature decreases. In both materials, a density of approximately 140 kg/m^3 allows for a higher strength than at a density of approximately 64 kg/m^3. High-density PVC-type foam is approximately 30% weaker than PUF at −110 °C or lower temperatures, whereas PVC-type foam is approximately 40% stronger than PUF at all temperatures.
- Overall, the impact test conveys that the initial peak forces of PVC-type foam are higher than those of PUF, while the absorption energy ratios are similar. Hence, PVC-type foam offers tangible advantages over PUFs in terms of impact resistance.

This study's results present an opportunity for using PVC-type foam, which satisfied a high-strength requirement, even at low temperature. Additionally, PVC-type foam offers substantial strength with low density. This serves the trend of increasing CCS wall thickness. Although PVC-type foam thermal conductivity was higher than that of PUF, all mechanical strength aspects (tensile, compressive, and impact) were stronger in the former. Consequently, given that PVC-type foam provides higher strength, it represents a more advantageous alternative insulation material/panel (for cryogenic temperatures). Furthermore, it can fulfill not only CCS needs but also other applications that require low density, high strength, and cryogenic environments.

Author Contributions: Conceptualization, D.-H.K. and J.-H.K.; project administration, J.-H.K., M.K., S.-K.K. and J.-M.L.; methodology, D.-H.K. and H.-T.K.; software, J.-D.K. and C.U.; validation, S.-K.K. and M.K.; formal analysis, D.-H.K. and H.-T.K.; investigation, H.-T.K., J.-M.L. and J.-H.K.; resources, J.-D.K. and C.U.; data curation, D.-H.K.; writing—original draft preparation, D.-H.K.; writing—review and editing, J.-M.L. and J.-H.K.; funding acquisition, J.-M.L.; supervision, J.-M.L. All authors have read and agreed to the published version of the manuscript.

Funding: This work was supported by the Korea Institute of Energy Technology Evaluation and Planning (KETEP) and the Ministry of Trade, Industry & Energy (MOTIE) of the Republic of Korea (20224000000090). This work was supported by the R&D Platform Establishment of Eco-Friendly Hydrogen Propulsion Ship Program (No. 20006632) funded by the Ministry of Trade, Industry & Energy (MOTIE, Republic of Korea). This work was supported by the Development and Demonstration of Eco-friendly Ocean Clean-up Vessel (RS-2022-00154674, Development of LNG-Hydrogen Fuel cell based Hybrid System for Ocean Clean-up Vessel) funded by the Ministry of Trade, Industry & Energy (MOTIE, Republic of Korea).

Institutional Review Board Statement: Not applicable.

Data Availability Statement: The data presented in this study are available on request from the corresponding author.

Conflicts of Interest: The authors declare no conflict of interest.

References

1. Galdo, M. Marine Engines Performance and Emissions. *J. Mar. Sci. Eng.* **2021**, *9*, 280. [CrossRef]
2. Lindstad, H.E.; Rehn, C.F.; Eskeland, G.S. Sulphur abatement globally in maritime shipping. *Transp. Res. Part D* **2017**, *57*, 303–313. [CrossRef]
3. Slaughter, A.; Ray, S.; Shattuck, T. *International Maritime Organization (IMO) 2020 Strategies in a Non-Compliant World*; Deloitte: London, UK, 2019.
4. Kumar, S.; Kwon, H.-T.; Choi, K.-H.; Lim, W.; Cho, J.H.; Tak, K.; Moon, I. LNG: An eco-friendly cryogenic fuel for sustainable development. *Appl. Energy* **2011**, *88*, 4264–4273. [CrossRef]
5. BP. *Statistical Review of World Energy Globally Consistent Data on World Energy Markets*; BP: London, UK, 2020; p. 66.
6. Choe, Y.-R.; Kim, J.-H.; Kim, J.-M.; Park, S.; Park, K.H.; Lee, J.-M. Evaluation of Cryogenic Compressive Strength of Divinycell of NO 96-type LNG Insulation System. *J. Ocean Eng. Technol.* **2016**, *30*, 349–355. [CrossRef]
7. Lloyd's Register. *List of Lloyd's Register Approved Material/Components for GTT Mk III & Mk V Membrane Containment System*; Lloyd's Register Group Services Ltd.: London, UK, 2019; pp. 1–26.
8. Diab. Divinycell CY SI. 2020. Available online: https://www.diabgroup.com/products-services/divinycell-pvc/divinycell-h/ (accessed on 14 November 2022).
9. Dnv, G.L. *Class Guideline Liquefied Gas Carriers with Membrane Tanks*; American Bureau of Shipping: Houston, TX, USA, 2020.
10. Lee, Y.; Choi, S.; Choe, K.; Kim, S. Physical and mechanical characteristics of polyurethane foam insulators blown by HFC. In Proceedings of the 15th International Offshore and Polar Engineering Conference, Seoul, Republic of Korea, 19–24 June 2005.
11. Grieve, P.; GST. *A New Generation of LNG Membrane-Type Land Storage Tank*; Gaztransport & Technigaz: Chevreuse, France, 2010; Available online: https://gtt.fr/sites/default/files/2010_lng16-gst-new-generation-lng-membrane-type-storage-tank-2012-04-11.pdf (accessed on 14 November 2022).
12. Jeong, H.; Shim, W.J. Calculation of Boil-Off Gas (BOG) Generation of KC-1 Membrane LNG Tank with High Density Rigid Polyurethane Foam by Numerical Analysis. *Pol. Marit. Res.* **2017**, *24*, 100–114. [CrossRef]
13. Lee, J.M.; Kim, M.H. Impact Strength Assessment of LNG Carrier Insulation System. *Key Eng. Mater.* **2006**, *326–328*, 1527–1530.
14. Chun, M.S.; Kim, M.H.; Kim, W.S.; Kim, S.H.; Lee, J.-M. Experimental investigation on the impact behavior of membrane-type LNG carrier insulation system. *J. Loss Prev. Process Ind.* **2009**, *22*, 901–907. [CrossRef]
15. Lu, J.; Xu, S.; Deng, J.; Wu, W.; Wu, H.; Yang, Z. Numerical prediction of temperature field for cargo containment system (CCS) of LNG carriers during pre-cooling operations. *J. Nat. Gas Sci. Eng.* **2016**, *29*, 382–391. [CrossRef]
16. Kim, M.H.; Kil, Y.P.; Lee, J.M.; Chun, M.S.; Suh, Y.S.; Kim, W.S.; Noh, B.J.; Yoon, J.H.; Kim, M.S.; Urm, H.S. Cryogenic Fatigue Strength Assessment for MARK-III Insulation System of LNG Carriers. *J. Offshore Mech. Arct. Eng.* **2011**, *133*, 041401. [CrossRef]
17. Iii, M.; Density, H. GTT Inside 3. 2015. Available online: https://gtt.fr/news/news-newsletter-gtt-inside-3-january-2015 (accessed on 14 November 2022).
18. Hiroshi, T.; Hirotomo, O.; Toshinori, I.; Satoshi, M. New LPG Carrier Adopting Highly Reliable Cargo Tank-IMO Tank Type B-. *Mitsubishi Heavy Ind. Tech. Rev.* **2013**, *50*, 12–17.
19. Park, S.-B.; Choi, S.-W.; Kim, J.-H.; Bang, C.-S.; Lee, J.-M. Effect of the blowing agent on the low-temperature mechanical properties of CO_2- and HFC-245fa-blown glass-fiber-reinforced polyurethane foams. *Compos. Part B Eng.* **2016**, *93*, 317–327. [CrossRef]
20. Son, Y.; Kim, J.; Choi, S.; Her, N.Y.; Lee, J. Synthesis of Polyurethane Foam Considering Mixture Blowing Agents for Application to Cryogenic Environments. *Macromol. Mater. Eng.* **2019**, *304*, 1900294. [CrossRef]
21. Marşavina, L.; Linul, E. Fracture toughness of rigid polymeric foams: A review. *Fatigue Fract. Eng. Mater. Struct.* **2020**, *43*, 2483–2514. [CrossRef]
22. Arvidson, J.M.; Sparks, L.L. *Low Temperature Mechanical Prop-Erties of a Polyurethane Foam*; National Bureau of Standards, U.S. Department of Commerce: Denver, CO, USA, 1981.
23. Farshidi, A.; Berggreen, C.; Carlsson, L.A. Low temperature mixed-mode debond fracture and fatigue characterisation of foam core sandwich. *J. Sandw. Struct. Mater.* **2018**, *22*, 1039–1054. [CrossRef]

24. Demharter, A. Polyurethane rigid foam, a proven thermalinsulating material for applications between +130 °C and −196 °C. *Cryogenics* **1998**, *38*, 113–117. [CrossRef]
25. Yakushin, V.A.; Zhmud, N.P.; Stirna, U.K. Physicomechanical characteristics of spray-on rigid polyurethane foams at normaland low temperatures. *Mech. Compos. Mater.* **2002**, *38*, 273–280. [CrossRef]
26. Stirna, U.; Beverte, I.; Yakushin, V.; Cabulis, U. Mechanical properties of rigid polyurethane foams at room and cryogenic temperatures. *J. Cell. Plast.* **2011**, *47*, 337–355. [CrossRef]
27. Denay, A.G.; Castagnet, S.; Roy, A.; Alise, G.; Thenard, N. Compres-sion behavior of glass-fiber-reinforced and pure polyurethanefoams at negative temperatures down to cryogenic ones. *J. Cell. Plast.* **2013**, *49*, 209–222. [CrossRef]
28. Yakushin, V.; Cabulis, U.; Sevastyanova, I. Effect of filler type on the properties of rigid polyurethane foams at a cryogenic temperature. *Mech. Compos. Mater.* **2015**, *51*, 447–454. [CrossRef]
29. Hwang, B.-K.; Kim, S.-K.; Kim, J.-H.; Lee, J.-M. Dynamic compressive behavior of rigid polyurethane foam with various densities under different temperatures. *Int. J. Mech. Sci.* **2020**, *180*, 105657. [CrossRef]
30. Lee, Y.; Choi, S.; Choi, G. Characteristics of Rigid Polyurethane Foams Blown by HFCs for LNG Storage Tank. *J. Korean Inst. Gas* **2005**, *9*, 16–20.
31. Kara, M.; Kırıcı, M.; Tatar, A.C.; Avcı, A. Impact behavior of carbon fiber/epoxy composite tubes reinforced with multi-walled carbon nanotubes at cryogenic environment. *Compos. Part B Eng.* **2018**, *145*, 145–154. [CrossRef]
32. Saint-Michel, F.; Chazeau, L.; Cavaillé, J.-Y.; Chabert, E. Mechanical properties of high density polyurethane foams: I. Effect of the density. *Compos. Sci. Technol.* **2006**, *66*, 2700–2708. [CrossRef]
33. Saha, M.; Mahfuz, H.; Chakravarty, U.; Uddin, M.; Kabir, M.E.; Jeelani, S. Effect of density, microstructure, and strain rate on compression behavior of polymeric foams. *Mater. Sci. Eng. A* **2005**, *406*, 328–336. [CrossRef]
34. Gibson, L.J.; Ashby, M.F. *Cellular Solids: Structure & Properties*, 1st ed.; Pergamon Press: Oxford, UK, 1988.
35. Park, J.; Kim, H.; Kim, J.; Kim, J.; Kim, S.; Lee, J. Eco-friendly blowing agent, HCFO-1233zd, for the synthesis of polyurethane foam as cryogenic insulation. *J. Appl. Polym. Sci.* **2021**, *139*, 51492. [CrossRef]
36. Goldstein, J.I.; Newbury, D.E.; Michael, J.R.; Ritchie, N.W.M.; Scott, J.H.J.; Joy, D.C. *Scanning Electron Microscopy and X-ray Microanalysis*; Springer: New York, NY, USA, 2017. [CrossRef]
37. Yao, H.; Pang, Y.; Liu, X.; Qu, J. Experimental Study of the Dynamic and Static Compression Mechanical Properties of Closed-Cell PVC Foams. *Polymers* **2022**, *14*, 3522. [CrossRef]
38. Thirumal, M.; Khastgir, D.; Singha, N.K.; Manjunath, B.; Naik, Y. Mechanical, Morphological and Thermal Properties of Rigid Polyurethane Foam: Effect of the Fillers. *Cell. Polym.* **2007**, *26*, 245–259. [CrossRef]
39. Tang, Y.; Zhang, W.; Jiang, X.; Zhao, J.; Xie, W.; Chen, T. Experimental investigations on phenomenological constitutive model of closed-cell PVC foam considering the effects of density, strain rate and anisotropy. *Compos. Part B Eng.* **2022**, *238*, 109885. [CrossRef]
40. Raje, A.; Buhr, K.; Koll, J.; Lillepärg, J.; Abetz, V.; Handge, U.A. Open-Celled Foams of Polyethersulfone/Poly(N-vinylpyrrolidone) Blends for Ultrafiltration Applications. *Polymers* **2022**, *14*, 1177. [CrossRef]
41. Maiti, S.K.; Gibson, L.J.; Ashby, M.F. Deformation and energy absorption diagrams for cellular solids. *Acta Met.* **1984**, *32*, 1963–1975. [CrossRef]
42. Jarfelt, U.; Ramnäs, O. Thermal conductivity of polyurethane foam—Best performance Thermal conductivity of polyurethane foam Best performance. In Proceedings of the 10th International Symposium on District Heating and Cooling, Reykjavik, Iceland, 3–5 September 2006; p. 12.
43. Choe, H.; Choi, Y.; Kim, J.H. Threshold cell diameter for high thermal insulation of water-blown rigid polyurethane foams. *J. Ind. Eng. Chem.* **2019**, *73*, 344–350. [CrossRef]
44. Kabir, E.; Saha, M.; Jeelani, S. Tensile and fracture behavior of polymer foams. *Mater. Sci. Eng. A* **2006**, *429*, 225–235. [CrossRef]
45. Tu, Z.; Shim, V.; Lim, C. Plastic deformation modes in rigid polyurethane foam under static loading. *Int. J. Solids Struct.* **2001**, *38*, 9267–9279. [CrossRef]
46. Park, S.-B.; Lee, C.-S.; Choi, S.-W.; Kim, J.-H.; Bang, C.-S.; Lee, J.-M. Polymeric foams for cryogenic temperature application: Temperature range for non-recovery and brittle-fracture of microstructure. *Compos. Struct.* **2016**, *136*, 258–269. [CrossRef]
47. Zhang, Y.; Liu, Q.; He, Z.; Zong, Z.; Fang, J. Dynamic impact response of aluminum honeycombs filled with Expanded Polypropylene foam. *Compos. Part B Eng.* **2019**, *156*, 17–27. [CrossRef]
48. Banyay, G.; Shaltout, M.; Tiwari, H.; Mehta, B. Polymer and composite foam for hydrogen storage application. *J. Mater. Process Technol.* **2007**, *191*, 102–105. [CrossRef]

Disclaimer/Publisher's Note: The statements, opinions and data contained in all publications are solely those of the individual author(s) and contributor(s) and not of MDPI and/or the editor(s). MDPI and/or the editor(s) disclaim responsibility for any injury to people or property resulting from any ideas, methods, instructions or products referred to in the content.

Article

Controlled Deposition of Single-Walled Carbon Nanotubes Doped Nanofibers Mats for Improving the Interlaminar Properties of Glass Fiber Hybrid Composites

Arif Muhammad [1], Mkhululi Ncube [2], Nithish Aravinth [3] and Jacob Muthu [3,*]

[1] Faculty of Engineering & Technology, International Islamic University, Islamabad 44000, Pakistan
[2] Faculty of Engineering and the Built Environment, University of the Witwatersrand, Johannesburg 2000, South Africa
[3] Faculty of Engineering & Applied Science, University of Regina, Regina, SK S4S 0A2, Canada
* Correspondence: jacob.muthu@uregina.ca

Abstract: The properties of glass fiber composites were improved by strengthening the interlaminar regions using electrospun nanofibers mats. However, the chaotic nature of the electrospinning process at the collector restricts the controlled deposition and alignment of nanofibers and limits the use of electrospun nanofibers as secondary reinforcements. Hence, auxiliary vertical electrodes were used, which drastically reduced the diameter of the nanofibers from 450 nm to 150 nm and also improved the alignment of nanofibers. The aligned nanofibers were then used for doping the functionalized single-walled carbon nanotubes (f-SWCNTs) with nanofibers, which controlled the inherent issues associated with SWCNTs such as agglomeration, poor dispersion, and alignment. This process produced f-SWCNTs doped nanofiber mats. A series of tensile, three-point flexural, and Charpy impact tests showed that 30 vol% glass fiber composites reinforced with 0.5 wt% of randomly oriented nanofiber (RONFs) mats improved the properties of the hybrid composites compared to 0.1 wt%, 0.2 wt%, and 1 wt% RONFs mats reinforced glass fiber hybrid composites. The increase in properties for 0.5 wt% composites was attributed to the higher specific surface area and resistance to the relative slip within the interlaminar regions. The 0.5 wt% RONFs were then used to produce 0.5 wt% of continuous-aligned nanofiber (CANFs) mats, which showed improved mechanical properties compared to 0.5 wt% randomly oriented nanofiber (RONFs) mats reinforced hybrid composites. The CANFs mats with reduced diameter increased the tensile strength, flexural strength, and impact resistance by 4.71%, 17.19%, and 20.53%, respectively, as compared to the random nanofiber mats. The increase in properties could be attributed to the reduced diameter, controlled deposition, and alignment of the nanofibers. Further, the highest increase in mechanical properties was achieved by the addition of f-SWCNTs doped CANFs mats strengthened hybrid composites, and the increase was 30.34% for tensile strength, 30.18% for flexural strength, and 132.29% for impact resistance, respectively. This improvement in properties was made possible by orderly alignments of f-SWCNTs within the nanofibers. The SEM images further confirmed that auxiliary vertical electrodes (AVEs) reduced the diameter, improved the alignment and molecular orientation of the nanofibers, and thus helped to reinforce the f-SWCNTs within the nanomats, which improved the properties of the glass hybrid composites.

Keywords: hybrid composites; aligned nanomats; functionalized SWCNTs; electrospinning; interlaminar region

Citation: Muhammad, A.; Ncube, M.; Aravinth, N.; Muthu, J. Controlled Deposition of Single-Walled Carbon Nanotubes Doped Nanofibers Mats for Improving the Interlaminar Properties of Glass Fiber Hybrid Composites. *Polymers* 2023, 15, 957. https://doi.org/10.3390/polym15040957

Academic Editor: Antonino Alessi

Received: 21 January 2023
Revised: 8 February 2023
Accepted: 10 February 2023
Published: 15 February 2023

Copyright: © 2023 by the authors. Licensee MDPI, Basel, Switzerland. This article is an open access article distributed under the terms and conditions of the Creative Commons Attribution (CC BY) license (https://creativecommons.org/licenses/by/4.0/).

1. Introduction

Glass fiber composites have a major role in various engineering applications due to their design flexibility, ease of manufacturing, and low weight compared to conventional materials such as steel [1,2]. However, the failure at the matrix-rich interlaminar regions due to poor mechanical properties is a cause of concern, and even with the higher volume fraction of glass fibers (60%), these matrix-rich regions exist [3]. Therefore, researchers are focused on improving the interlaminar regions by adding secondary reinforcements such as nanofillers [4–6], and among them, polymeric nanofibers are gaining greater importance due to their ease of manufacturing with unique mechanical properties [6].

Research works [7,8] show that continuous nanofibers have a higher molecular orientation with a better degree of crystallinity and, thus, they improve the properties of the interlaminar regions. Wang et al. [9] cited that continuous nanofibers could improve the mechanical properties along a particular direction and promote load-bearing capabilities. These results expounded the innovative ideas in manufacturing aligned and continuous nanofibers in the form of nanomats. These nanomats could develop a bridging mechanism within the interlaminar regions and be suitable interlaminar reinforcements for glass fiber hybrid composites. The continuous-aligned nanofibers (CANFs) mats will be thin, light in weight, and will not increase the weight of final composites. Moreover, they can reduce the length scale mismatch between the macroscale fibers and the polymer matrix molecular chains. In addition, these nanomats could be collected directly over the fiber mats and thus reduce the processing time/cost.

Several studies focus on using various nanoparticles prepared as mats for reinforcing composites such as graphene-embedded paper for electromagnetic interference shielding (EMI) [10], graphene/SiCnw nanostrctured films for improving the mechanical and thermal properties of carbon fiber composites [11]. Among them, the single-walled carbon nanotubes (SWCNTs), though they are comparatively expensive, have attracted researchers due to their exceptional mechanical properties with higher specific surface area [12]. However, the inherent issues, such as agglomeration, alignment, and hydrophobic nature, limit their potential as secondary reinforcements, which necessitated the innovative research ideas for using SWCNTs as interlaminar reinforcements [13]. One such direction could be to develop continuous-aligned nanofibers mats doped with SWCNTs.

Several techniques are being used for producing polymeric nanofibers, and among them, electrospinning is considered simple and versatile. However, the polymeric jets formed during electrospinning have two distinct phases: near field, and far field [14]. In the near field, the jet takes a straight path, and in the far field, it starts whipping into a complex path due to electrical, gravitational, and rheological forces. Though the far field yields nanosized fibers [15], the whipping leads to randomly oriented nanofibers, which defeats the objective of producing continuous and aligned nanofibers. Moreover, the nanosized fibers are vital for aligning and dispersing SWCNTs into the nanofibers mats. To overcome the whipping instability, researchers have used thin static collectors [16] or high-speed rotating collectors [17]. However, based on our experience, a wide rotating drum collector with auxiliary vertical electrodes (AVEs) seems to be the most suitable solution for controlling the whipping effect and also for improving the alignment of nanofibers within the nanomats. Moreover, AVEs have better control over the whipping instability at the far field, as confirmed by other researchers [18].

Hence, the focus of this study is to utilize the electrospinning setup with AVEs for producing CANFs mats and functionalized SWCNTs (f-SWCNTs) doped CANFs mats. The SWCNTs will be functionalized to improve the interfacial adhesion with nanofibers. Lastly, the effect of CANFs and f-SWCNTs doped CANFs mats on the mechanical properties of glass fiber composites will be evaluated using tensile, flexural, and impact tests to characterize the interlaminar strengthening mechanism.

2. Materials and Methods

2.1. Materials

The polymer solution used in electrospinning was prepared by mixing Polyacrylonitrile powder (PAN: density 1.184 g/cm^3 at 25 °C) with N, N-dimethylformamide (DMF: 99% purity) supplied by Sigma Aldrich, Kempton Park, South Africa. The solution was stirred for 22 h at 50 °C until it became homogeneous. The composites were fabricated using Bisphenol-A epoxy (AMPREG 21 Resin) with woven E-glass mats supplied by AMT

Composites Pty. Ltd., Johannesburg, South Africa. The SWCNTs with 99% purity were used to dope CANFs for strengthening the interlaminar region. The SWCNTs and the required functionalization chemicals, such as Polyvinyl alcohol (PVA) and dimethyl sulfoxide (DMSO), were purchased from Sigma Aldrich, Kempton Park, South Africa. The physical and mechanical properties of Bisphenol-A epoxy, woven E-glass, and SWCNTs are given in Table 1.

Table 1. Properties of materials used—data collected from the suppliers.

Properties	Bisphenol-A Epoxy	Woven E-Glass Mats	SWCNTs
Tensile strength (MPa)	72.7–81.3	148.6–152.5	80,000–105,000
Tensile modulus (GPa)	3.3–4.3	7.2–8.3	~1000
Elongation at break (%)	1.12	2.4	-
Density at 23 °C (gm/cm^3)	1.148	2.6	1.7
Average diameter (nm)	-	-	0.83–1.3

2.2. Testing Procedure

The tensile, three-point flexural, and Charpy impact tests were carried out on the composite specimens as per ASTM D638-08: 2010, ASTM D256-06a: 2010, and ASTM D790-07: 2010, respectively. The specimen thickness was kept at 3.2 mm. The flexural test specimens were fabricated with the thickness to width ratio of 1:16, and the gauge length between the supports was kept at 63 mm. A 20 kN SHIMADZU tensile testing machine was used to conduct both tensile and flexural tests (Figure 1a,b, which were carried out at a cross-head speed of 2 mm/min. The Charpy impact tests were carried out on a Tinius Olsen impact pendulum as per the ISO 179-1 standard (Figure 1c. The breaking energy was used to determine the impact resistance of the composite specimens.

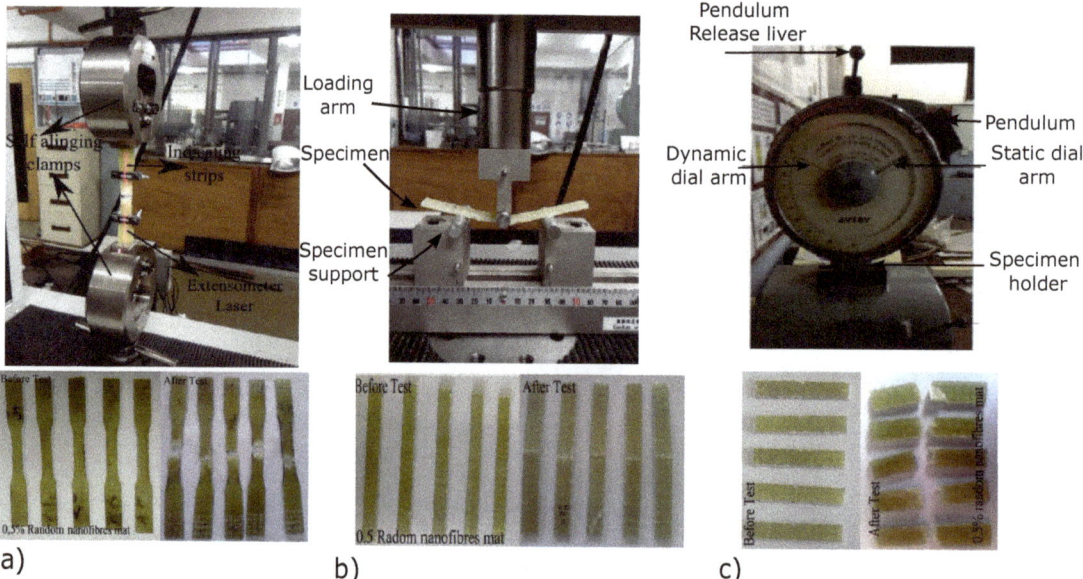

Figure 1. (**a**) Tensile testing setup and specimens. (**b**) Flexural testing setup and specimens. (**c**) Impact testing setup and specimens.

2.3. Morphological Characterization

Morphological characterization of nanofibers and hybrid composite specimens was carried out using scanning electron microscopy (SEM-Nova 600 Nanolab and Quanta 400 FEG from FEI Company, Hillsboro, OR, USA) at the voltage range of 5 KV to 30 KV. Before analysis, the specimens were coated with a thin layer of carbon and a 15 nm layer of gold–palladium (Au/Pd) in a ratio of (60:40) in an EMITECH K950X evaporator and EMITECH K550X sputter coater, respectively. The specimens were then attached to 1 cm diameter stubs with DAG 580 colloidal graphite. The FTIR analysis was conducted on a Bruker Vector 22 FTIR spectrometer using 64 sample scans and 32 background scans. The Raman analysis was conducted using a Horiba Jobin-Yvon Lab RAM HR Raman spectrometer with a 514.5 nm line of an argon-ion laser as the excitation source.

2.4. SWCNTs Functionalization

The SWCNTs were functionalized using the Friedel–Crafts alkylation (FCA) process. The FCA process was carried out by mixing SWCNTs (0.4 g) and PVA (4 g) with a 40 mL DMSO solution in a 250 mL flask using a magnetic stirrer. The flask was then immersed in an oil bath for gradual heating to 90 °C in the presence of nitrogen. As the heating promotes the chemical process, a catalyst of aluminum chloride ($ALCl_3$) was added for grafting the PVA chains onto the surface of the SWCNTs. During the chemical process, the OH groups from PVA reacted with $ALCl_3$ and produced positive carbocations, ($CH3^+$ ions : electrophiles), and negative $ALCl_2^-$ ions, respectively. The electrophiles were then attracted to the SWCNTs' hexagonal ring due to the CNT's surface delocalized electron (nucleophilic) sites, and thus attached the PVA chains onto the SWCNTs. After 20 h, the chemical reaction was aborted by adding 100 mL of methanol/hydrochloric acid mixture (volumetric ratio 1:1). The solution was then centrifuged at 3500 rpm for 15 min to precipitate the functionalized SWCNTs (f- SWCNTs). The f-SWCNTs were washed, filtered, and dried in a furnace for 3 h at 70°.

The Raman spectra were used to analyze the functionalization results of both pristine (p-SWCNTs) and functionalized SWCNTs (f-SWCNTs) and are shown in Figure 2i. The two peaks at the wavelengths of 1300–1400 cm^{-1} and 1500–1600 cm^{-1} were for the disorder (D band) and graphite band (G band), respectively. These bands represent two different vibrational modes of SWCNTs. The D band represents the crystal disorder, such as sidewall defects (pentagons or heptagons) or Sp3 carbon hybridization, and the G band was for the Sp2 carbon hybridization. The ratio of D band to G band (I_D/I_G) increases with the increasing disorder [19] and defines the structural changes of the SWCNTs wall due to functionalization [20]. In the current work, the I_D/I_G ratio was measured at 0.1025 and 0.1674 for p-SWCNTs and f-SWCNTs, respectively. The higher ratio of 0.1674 for f-SWCNTs was due to the carbocations attachment to the SWCNTs walls and the conversion of Sp2 to Sp3 hybridized carbons. This confirmed the PVA grafting to the SWCNTs wall during the FCA process [21].

Figure 2ii shows the FTIR spectrum for the range of 4000 to 500 cm^{-1} of both p-SWCNTs and f-SWCNTs. By analyzing the f-SWCNTs spectrum, the peak at 1028 cm^{-1} showed the C–O stretching, while the peak from 1100 to 1600 cm^{-1} confirmed the aromatic structures. The peaks corresponding to –CH and –OH bonds were obtained at 2900 cm^{-1} and 3200–3400 cm^{-1}, respectively. These peaks confirm that the PVA chains are attached to the surface of SWCNTs.

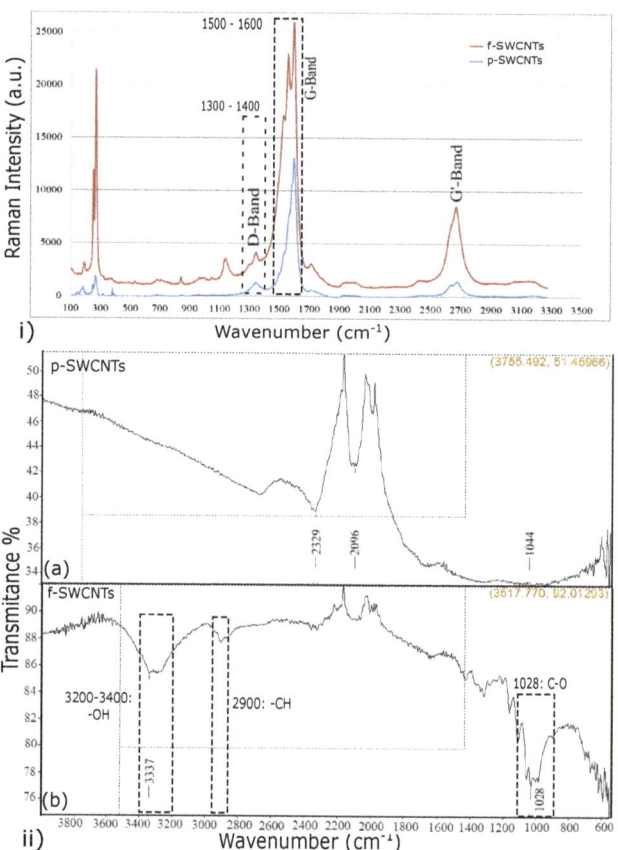

Figure 2. (**i**) Raman spectra of p-SWCNTs and f-SWCNTs. (**ii**) FTIR spectra of (**a**) p-SWCNTs and (**b**) f-SWCNTs.

2.5. Electrospinning of Aligned and Doped Nanofibers Mats

The electrospinning setup considered for this research consists of a DC voltage supply, NE-1600 syringe pump, and a grounded rotating drum collector (diameter 120 mm and width 210 mm), respectively. The electrospun nanomats were collected over the glass fiber mats attached to the rotating drum collector. The details of the complete setup with the spinning parameters were presented in our published manuscript [22] and the readers are encouraged to refer to the paper for further details. Following our previous experimental results, three electrospinning parameters, i.e., applied voltage, the distance between electrodes, and PAN concentration, were investigated in this work. In addition, auxiliary vertical electrodes (AVEs) were placed between the spinneret and collector. The range of tested and the optimized values of the spinning parameters are given in Table 2, which produced the nanofiber mats with fiber diameters in the range of 120–150 nm.

Table 2. Spinning parameter values with AVEs.

Electrospinning Parameters	Range of Tested Values	Optimized Values
Concentration (%)	8–8.8	8.1
Flow rate (mL/h)	0.12–0.50	0.35
Applied voltage (DC-KV)	0–30	20
Collector to spinneret distance (cm)	5–35	25

To analyze the effect of AVEs on the whipping effect, an electric field simulation software was used to simulate the electrostatic forces. The software developed the electrostatic forces as electric field lines (green lines) between the spinneret and the collector. The positive electrode (the spinneret) is red, while the blue color refers to the negative electrode. The simulation shows that without AVEs, the electric field lines are spread over the entire width of the collector. As the electric fields were not focused on the collector, the whipping effect deposited the nanofibers randomly over the whole width of the collector (Figure 3a). However, after introducing AVEs (Figure 3b), the simulation shows that the electric field lines were narrowed down towards the center of the collector (nanofiber driving zone) by the AVEs electric field. This phenomenon reduced the whipping effect and aligned the nanofibers perpendicular to the collector axis. Figure 4 shows the camera images and scanning electron microscope (SEM) images of electrospun nanofibers mats produced without (Figure 4a) and with (Figure 4b) AVEs. As it can be seen, the AVEs significantly improved the alignment of nanofibers and further reduced the diameter of the nanofibers to 150 nm, which is 1/3 smaller than the nanofibers produced without AVEs. This confirms that the addition of AVEs significantly refined the nanofiber morphologies.

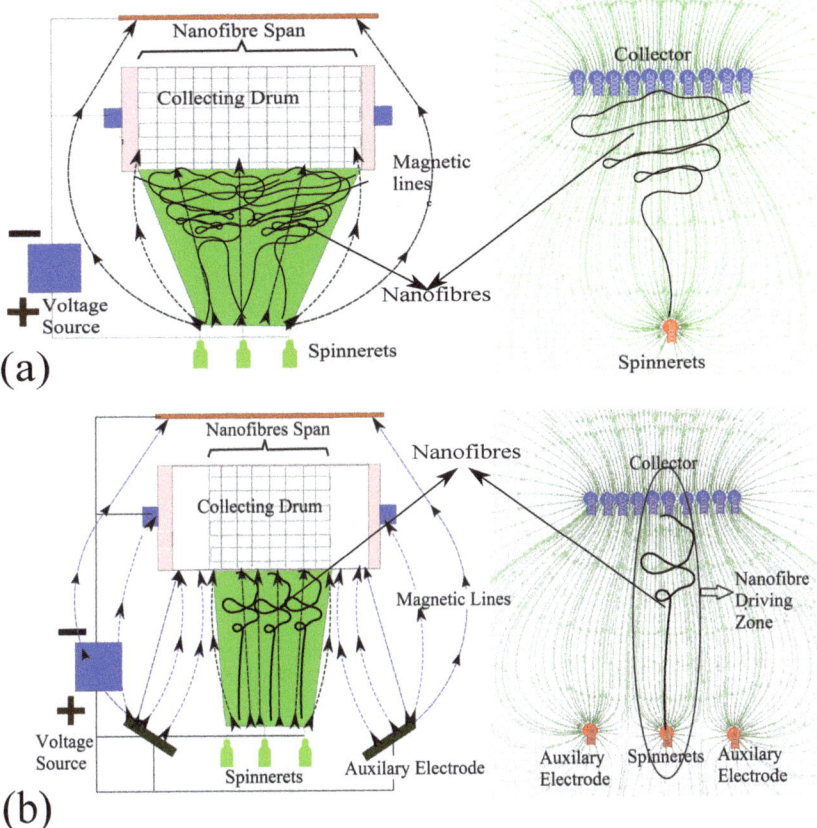

Figure 3. Electric field simulation software analysis (**a**) without AVEs; (**b**) with AVEs.

The electrospinning setup was used to produce the required nanomats. At first, the randomly oriented PAN nanofiber (RONFs) mats were produced without AVEs for the weight fractions of 0.1%, 0.2%, 0.5%, and 1%, respectively. Based on the experimental results, the weight fraction of 0.5 wt% PAN nanofibers were selected to manufacture continuous-aligned nanofibers (CANFs) mats using AVEs. Finally, the 0.5 wt% CANFs mats were doped with 0.25 wt% f-SWCNTs to produce f-SWCNTs doped CANFs mats. For doping, only 0.25 wt% f-SWCNTs were selected to minimize the density effect following the results from Pilehrood et al. [23]. Before spinning the nanomats, the required weight fractions of PAN-DMF solution with and without f-SWCNTs were prepared and electrospun over the glass fiber mats.

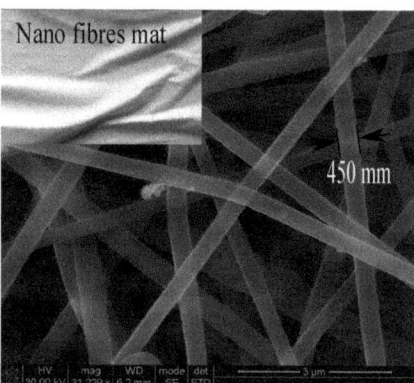

(a) Random Nanofibres Mat

(b) Aligned Nanofibres Mat

Figure 4. Electrospun nanofibers mats (**a**) without AVEs; (**b**) with AVEs.

2.6. Composites Manufacturing Process

The composites were fabricated using a vacuum-assisted resin transfer molding (VARTM) method. Following our previous work [4], 32 vol% glass fiber composites were selected as a reference for producing nanomats strengthened hybrid composites. Considering the manuscript length, the details of the glass fiber composites experimental results are not included, and the mechanical properties for the 32 vol% glass fiber composites are given in Table 3.

The VARTM mold consists of a female square steel ring, and upper and lower male plates, and these were assembled with the alternative layers of woven E-glass fiber mats coated with the RONFs, CANFs, and f-SWCNTs doped CANFs mats. Then, the resin-hardener mixer was placed at the mold suction port and the vacuum pressure of 80 psi was maintained until the resin filled the mold. Finally, the resin was allowed to cure for 24 h at room temperature. Figure 5 shows the manufacturing process of nanomats strengthened hybrid composites specimens, which were machined into tensile, flexural, and impact test specimens.

Table 3. Properties of 32 vol% glass fiber composites.

	Tensile Strength (MPa)	Elastic Modulus (GPa)	Flexural Strength (MPa)	Flexural Modulus (GPa)	Impact Resistance (KJ/m^2)
32 vol% glass fiber	156.52	11.77	242.2	9.58	160.18

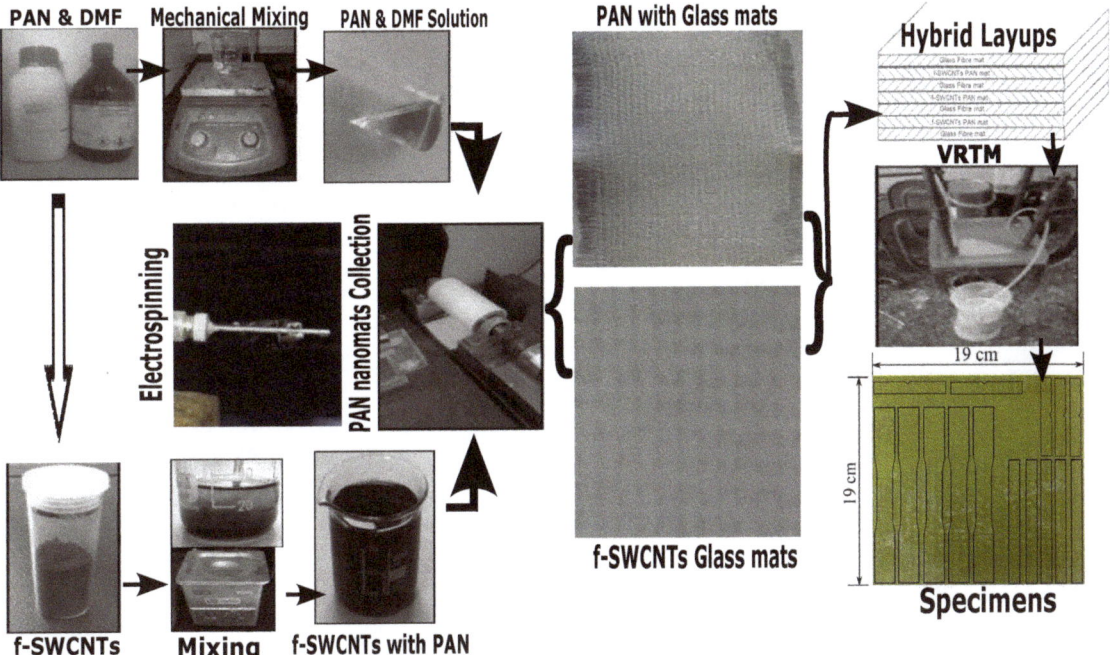

Figure 5. Fabrication process of glass fiber hybrid composites preform.

3. Results and Discussion

The discussions of the results are divided into two sections. The first section will focus on analyzing the results from RONFs mats strengthened hybrid composite specimens, and the CANFs mats and f-SWCNTs doped CANFs mats strengthened hybrid composites will be discussed in the second section.

3.1. RONFs Mats Strengthened Hybrid Composites

Figure 6 shows the mechanical properties of RONFs mats strengthened hybrid composites, and it uses the 32 vol% glass fiber composites as reference. The initial addition of 0.1 wt% of mats showed a scanty increase in properties compared to the 32 vol% glass fiber composites. Further addition of 0.5 wt% RONF mats showed an increase in tensile strength from 156.52 to 251.51 MPa, elastic modulus from 11.77 to 14.36 GPa, flexural strength from 242.2 to 404.01 MPa, flexural modulus from 9.58 GPa to 14.58 GPa, and impact resistance from 160.18 KJ/m^2 to 193.112 KJ/m^2, respectively. The statistical parameters for the RONFs strengthened glass hybrid composites are presented in Table 4.

The increase in the properties could be attributed to the reinforcing effect of RONFs mats due to their higher specific surface area, which could have increased the contact area within interlaminar regions. The infused RONFs mats provided more resistance to the relative slippage of the matrix over the glass fiber mats while transferring the applied load (refer to Figure 7a) and thus improved the mechanical properties. In addition, the geometrical web network of RONFs mats provided bridging between the glass fibers and the matrix, as shown in Figure 7b, and thus increased the resistance to the applied load and improved the properties.

Table 4. Statistical parameters of the RONFs strengthened glass fiber composites.

	Tensile Strength (MPa)				Elastic Modulus (GPa)				Flexural Strength (MPa)				Flexural Modulus (GPa)				Impact Resistance (KJ/m^2)			
RONFs wt%	0.1	0.2	0.5	1	0.1	0.2	0.5	1	0.1	0.2	0.5	1	0.1	0.2	0.5	1	0.1	0.2	0.5	1
No of samples	5	5	5	5	5	5	5	5	5	5	5	5	5	5	5	5	5	5	5	5
Mean	194.47	213.56	251.51	161.58	11.79	12.09	14.36	10.79	283.77	311.70	404.01	282.62	11.55	13.18	14.58	11.01	168.82	181.40	193.11	192.02
Median	207.78	213.66	262.38	164.32	11.42	12.32	14.54	11.09	285.54	329.91	409.08	239.87	10.89	13.91	14.18	10.88	162.67	177.05	189.35	193.45

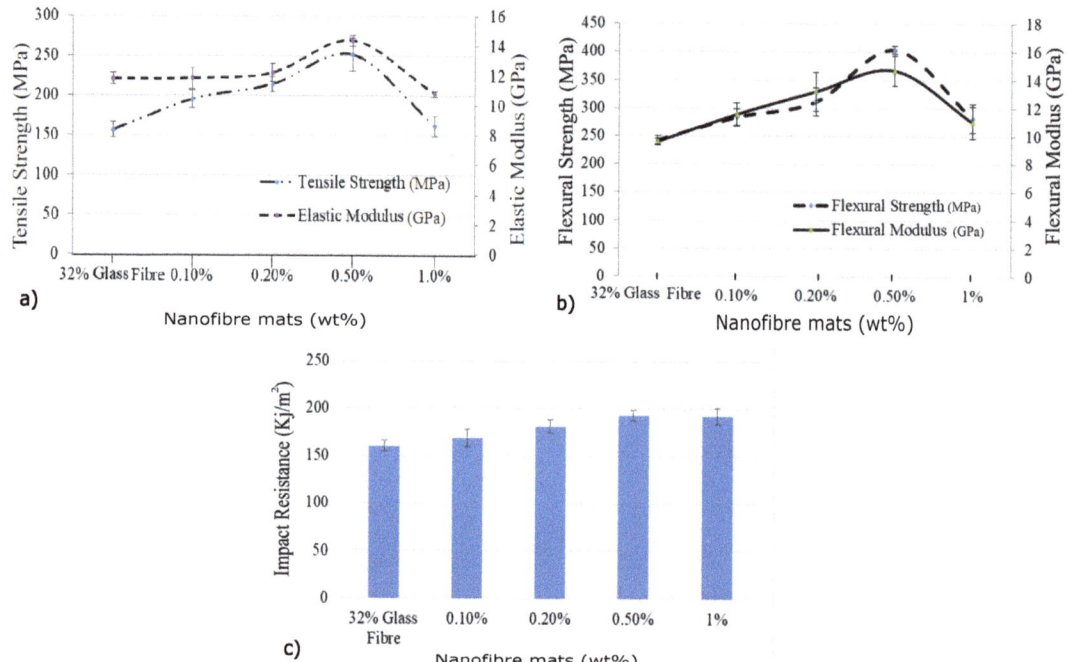

Figure 6. Properties of RONFs mats strengthened hybrid composite: (**a**) tensile properties; (**b**) flexural properties; (**c**) impact properties.

However, a similar effect was not observed for further increase in RONFs mats from 0.5 wt% to 1 wt%. This may be associated with the nanofibers agglomeration and poor penetration of resin into the nanofibers network due to the random orientation with higher weight fractions, which could have acted as stress concentration sites. Figure 8 provides evidence of agglomerated nanofibers within the interlaminar regions.

3.2. CANFs and f-SWCNTs Doped CANFs Mats Strengthened Hybrid Composites

Based on the above results, 0.5 wt% of nanomats were selected to manufacture 0.5 wt% CANFs mats and 0.25 wt% f-SWCNTs doped with 0.5 wt% CANFs mats strengthened glass fiber hybrid composites specimens to understand the effect of alignment and f-SWCNTs doping within the interlaminar regions. The experimental results used 0.5 wt% of RONF mats strengthened hybrid composites as a reference, and the results are presented below.

Figure 7. Fracture behavior of 0.5 wt% RONFs mats strengthened hybrid composite (**a**) Pulled out nanofibres (**b**) Bridging of nanofibres.

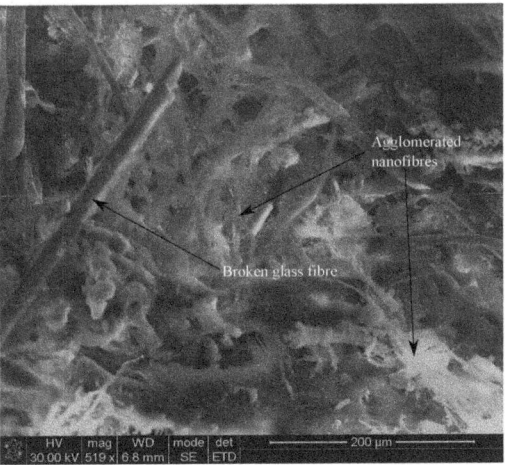

Figure 8. Agglomerated nanofibers of 1 wt% RONFs mats strengthened hybrid composite.

Mechanical Properties

The tensile, flexural, and impact properties of CANFs mats and f-SWCNTs doped CANFs mats strengthened hybrid composites are shown in Figure 9, and the properties values with percentage of increase are shown in Table 5.

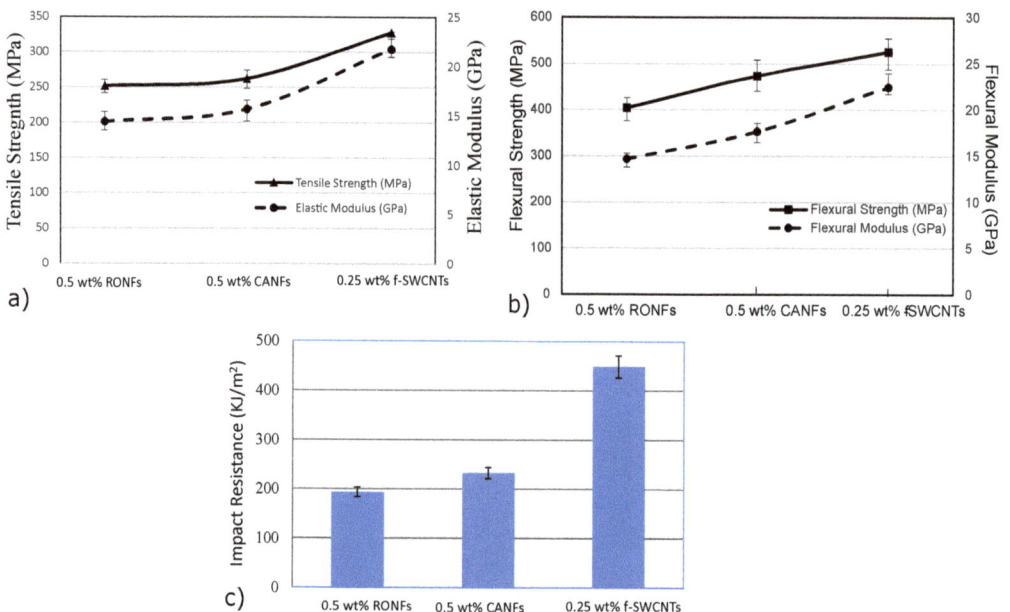

Figure 9. Mechanical properties of 0.5 wt% CANFs and 0.25 wt% f-SWCNTs doped with 0.5 wt% CANFs mats strengthened hybrid composites': (**a**) tensile properties; (**b**) flexural properties; (**c**) impact properties.

Table 5. RONF mats , CANF mats, and f-SWCNTS doped CANF mats hybrid composites properties and their increase.

Properties	0.5 wt% RONF Mats	0.5 wt% CANF Mats	% Increase	0.25wt% Doped with 0.5 wt% CANF Mats	% Increase
Tensile strength (MPa)	251.51	262.01	4.71	327.83	30.34
Elastic modulus (GPa)	14.36	15.63	8.84	27.72	93.04
Flexural strength (MPa)	404.01	473.44	17.19	525.96	30.18
Flexural modulus (GPa)	14.58	17.65	21.06	22.49	54.25
Impact resistance KJ/m^2	193.11	232.75	20.53	448.58	132.29

The phenomenal increase in mechanical properties was possible due to the reinforcement effect of continuous/aligned nanofibers along with uniformly dispersed f-SWCNTs doped CANFs mats within the interlaminar regions. The increase in mechanical properties can be attributed to the molecular orientations, reduction in nanofibers diameter, and the uniform dispersion of the aligned nanofibers network due to AVEs. Further, the reduction in nanofibers diameters (450 nm to 150 nm) provided an approximately threefold increase in nanofibers network within the interlaminar region.

Figure 10 shows the fractured surfaces of RONFs, CANFs, and f-SWCNTs doped CANFs mats strengthened hybrid composites. These images confirm that the RONFs mats (Figure 10a) accumulated nanofibers and created bigger voids as they were pulled out from the matrix. This led to an inefficient strengthening mechanism due to which the initiated cracks propagated without much resistance and resulted in poor mechanical properties. On the other hand, the CANFs mats with reduced diameter and improved distribution and orientation showed an orderly pull out of nanofibers and broken heads of nanofibers within the matrix (Figure 10c). This indicates that the bridging mechanism created by the nanomats networks acted as barriers for the propagating cracks, and, as such, the interlaminar regions were strengthened by the CANFs mats.

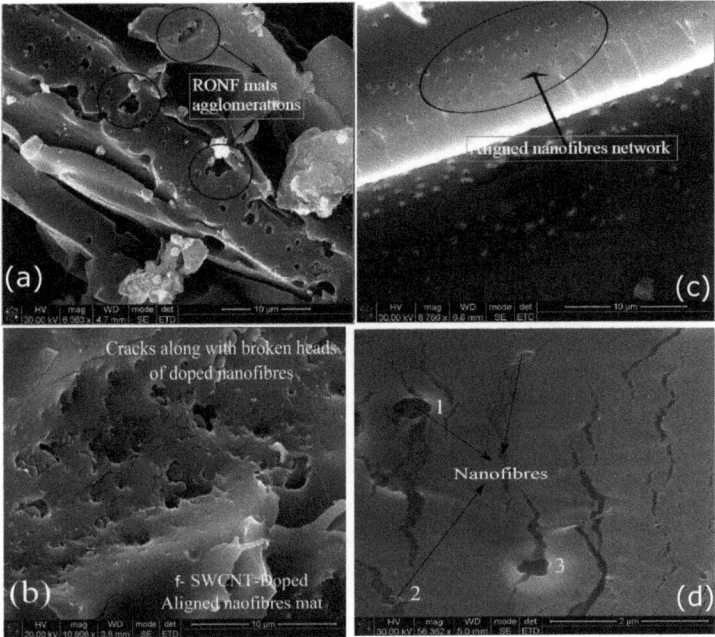

Figure 10. Fracture surfaces: (**a**) Agglomeration of RONFs mats strengthened hybrid composite tensile test specimen. (**b**) Tensile test specimen fracture surface of f-SWCNTs doped CANFs mats strengthened hybrid composite. (**c**) Fracture surface of the interlaminar regions with aligned nanomats networks of flexural test specimen. (**d**) Branched cracks within the interlaminar region due to nanofibers network.

Figure 10b shows the SEM images of the fractured surface of f-SWCNTs doped CANFs mats strengthened hybrid composite. The images show several propagating cracks around the f-SWCNTs doped nanofibers heads. This implies that the initiated cracks were arrested by the f-SWCNTs doped CANFs mats and resisted/diverted the propagating cracks. The resistance prevented the separation of the matrix due to the propagating cracks. This phenomenon is evident in Figure 10d, where the cracks were either stopped or diverted due to the presence of f-SWCNTs doped nanofibers (refer to sites 1, 2, and 3). At site 3, the crack traveled across the nanofiber after the f-SWCNTs doped nanofiber was pulled out, and in sites 1 and 2, the cracks propagated over the broken nanofiber heads or pulled out f-SWCNTs doped nanofibers. This confirmed that the increase in mechanical properties of hybrid composites is due to the resistive f-SWCNTs doped nanofibers mats network within the interlaminar regions.

As discussed before, the bridging mechanism within the multiscale hybrid composites can be seen from the SEM images, as shown in Figure 11a,b. The functionalization of SWCNTs provided improved the dispersion, and further strengthened the bonding between f-SWCNTs and nanofibers, which could have resulted in a stronger nanofibers network within the interlaminar regions. Further, the alignment provided the required bridging, which resisted the crack propagation and promoted load transfer within the interlaminar region. Thus, the f-SWCNTs doped aligned nanomats strengthened hybrid composite provided the highest mechanical properties.

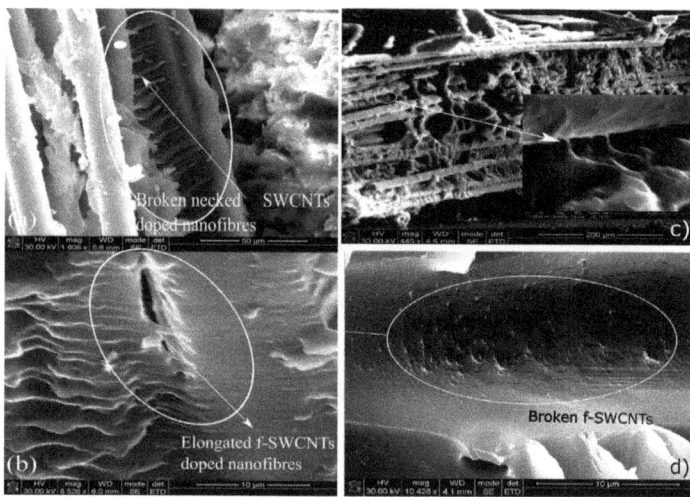

Figure 11. Fracture surfaces of 0.5 wt% nanofibers strengthened hybrid composites (**a**) Elongated f-SWCNTs doped CANFs mats in bridging. (**b**) Bridging of f-SWCNTs doped nanofibers. (**c**) Fractured surfaces of impact tested specimen. (**d**) Pulled and broken head of f-SWCNTs.

4. Conclusions and Recommendations

The continuously aligned nanofiber mats with and without f-SWCNTs were produced using an electrospinning setup with AVEs, which were then used to reinforce the interlaminar regions of glass fiber composites. The effect of aligned nanomats on the mechanical properties of glass fiber composites was seen from the experimental results; as such, both CANFs and f-SWCNTs doped CANFs produced hybrid composites with improved properties by strengthening the interlaminar properties. The addition of AVEs with the electrospinning reduced the nanofiber diameter and improved their alignments within the interlaminar regions. The reduced diameter and alignment of nanofibers helped to improve the molecular orientation and further helped to distribute f-SWCNTs uniformly into the glass fiber hybrid composites. The functionalization of SWCNTs promoted better dispersion of SWCNTs and improved the bonding between f-SWCNTs and nanofibers. These factors promoted better load transfer within the interlaminar regions and thus improved the mechanical properties of the glass fiber composites. The research could serve in different applications, such as electronics, filtration, etc., where aligned nanofibers with reduced diameter and improved molecular orientation are the essential requirements. The application of SWCNTs could also be extended to other research areas, such as smart textiles, transistors, information storage devices, batteries, EMI shielding, etc.

Author Contributions: Conceptualization and investigation: A.M. and M.N. Experiments and resources: A.M. and N.A. Writing—review and editing: A.M., N.A. and J.M. Supervision and funding acquisition: J.M. All authors have read and agreed to the published version of the manuscript.

Funding: This project was partially supported by the funding from Natural Sciences and Engineering Research Council of Canada (DDG-2020-00046) and National Research Foundation South Africa.

Institutional Review Board Statement: Not applicable.

Data Availability Statement: Data are contained within the article.

Conflicts of Interest: The authors declare no conflict of interest.

References

1. Masuelli, M. Introduction of Fibre-Reinforced Polymers—Polymers and Composites: Concepts, Properties and Processes. In *Fibre Reinforced Polymers—The Technology Applied for Concrete Repair*; InTechOpen: London, UK, 2013. [CrossRef]
2. Jung, S.M.; Yoon, G.H.; Lee, H.C. Chitosan nanoparticle/PCL nanofiber composite for wound dressing and drug delivery. *J. Biomater. Sci.* **2008**, *26*, 252–263. [CrossRef] [PubMed]
3. Fan, Z.; Santare, M.H.; Advani, S.G. Interlaminar shear strength of glass fiber reinforced epoxy composites enhanced with multiwalled carbon nano-tubes. *Compos. Part A Appl. Sci. Manuf.* **2008**, *39*, 540–554. [CrossRef]
4. Muthu, J.; Dendere, C. Functionalized multiwall carbon tubes strengthened GRP hybrid composites: Improved properties with optimum fiber content. *Compos. Part B* **2014**, *67*, 84–94. [CrossRef]
5. Ren, Z.; Gao, P.X. A review of helical nanostructures: Growth theories, synthesis strategies and properties. *Nanoscale* **2014**, *6*, 9366–9400. [CrossRef] [PubMed]
6. Rahmanian, S.; Thean, K.S.; Suraya, A.R.; Shazed, M.A.; Salleh, M.A.M.; Yusoff, H.M. Carbon and glass hierarchical fibers: Influence of carbon nanotubes on tensile, flexural and impact properties of short fiber reinforced composites. *Mater. Des.* **2013**, *43*, 10–16. [CrossRef]
7. Liao, H.; Wu, Y.; Wu, M.; Zhan, X.; Liu, H. Aligned electrospun cellulose fibersreinforced epoxy resin composite films with high visible light transmittance. *Cellulose* **2012**, *19*, 111–119. [CrossRef]
8. Karimi, S.; Staiger, M.P.; Buunk, N.; Fessard, A.; Tucker, N. Uniaxially aligned electrospun fibers for advanced nanocomposites based on a modelPVOH-epoxy system. *Compos. Part A Appl. Sci. Manuf.* **2016**, *81*, 214–221. [CrossRef]
9. Wang, G.; Yu, D.; Kelkar, A.D.; Zhang, L. Electrospun nanofiber: Emerging reinforcing filler in polymer matrix composite materials. *Prog. Polym. Sci.* **2017**, *75*, 73–107. [CrossRef]
10. Ud DinK, I.; Naresh, K.; Umer, R.; Khan, K.A.; Drzal, L.T.; Haq, M.; Cantwell, W.J. Processing and out-of-plane properties of composites with embedded graphene paper for EMI shielding applications. *Compos. Part A* **2020**, *134*, 105901–105913. [CrossRef]
11. Wang, F.; Wang, B.; Zhang, Y.; Zhao, F.; Qiu, Z.; Zhou, L.; Chen, S.; Shi, M.; Huang, Z. Enhanced thermal and mechanical properties of carbon fiber/epoxy composites interleaved with graphene/SiCnw nanostructured films. *Compos. Part A* **2022**, *162*, 107129–107141. [CrossRef]
12. Vajtai, R. *Springer Handbook of Nanomaterials*; Springer: Berlin/Heidelberg, Germany, 2013; pp. 105–135.
13. Ma, P.C; Siddiqui, N.A.; Marom, G.; Kim, J.K. Dispersion and functionalization of carbon nanotubes for polymer-based nanocomposites: A review. *Compos. Part A* **2010**, *41*, 1345–1367. [CrossRef]
14. Ramakrishna, S. *An Introduction to Electrospinning and Nanofibers*; World Scientific: Singapore, 2005; pp. 1–18.
15. Su, Z.; Ding, J.; Wei, G. Electrospinning: A facile technique for fabricating polymeric nanofibers doped with carbon nanotubes and metallic nanoparticles for sensor applications. *R. Soc. Chem.* **2014**, *4*, 52598–52610. [CrossRef]
16. Reneker, D.H; Yarin, A.L. Electrospinning jets and polymer nanofibers. *Polymer* **2008**, *49*, 2387–2425. [CrossRef]
17. Li, D.; Wang, Y.; Xia, Y. Electrospinning of polymeric and ceramic nanofbers as uniaxially aligned arrays. *Nano Lett.* **2003**, *3*, 1167–1171. [CrossRef]
18. Zhao, J.; Liu, H.; Xu, L. Preparation and formation mechanism of highly aligned electrospun nanofibers using a modified parallel electrode method. *Mater. Des.* **2016**, *90*, 1–6. [CrossRef]
19. Saito, R.; Dresselhaus, G.; Dresselhaus, M.S. *Physical Properties of Carbon Nanotubes*; Imperial College Press: London, UK, 1998.
20. Wepasnick, K.A.; Smith, B.A.; Bitter, J.L.; Howard Fairbrother, D. Chemical and structural characterization of carbon nanotube surfaces. *Anal. Bioanal. Chem.* **2010**, *396*, 1003–1014. [CrossRef] [PubMed]
21. Wu, X.L.; Liu, P. Poly(vinyl chloride)-grafted multi-walled carbon nanotubes via Friedel-Crafts alkylation. *Express Polym. Lett.* **2010**, *4*, 723–728. [CrossRef]
22. Muthu, S.D.J.; Bradley, P.; Jinasena, I.K.; Wegner, L.D. Electro spun nanomats strengthened glass fiber hybrid composites: Improved mechanical properties using continuous nanofibers. *Polym. Compos.* **2020**, *41*, 958–971. [CrossRef]
23. Pilehrood, M.K.; Heikkilä, P.; Harlin, A. Preparation of carbon nanotubes embedded in polyacrylonitrile (PAN). *Autex Res. J.* **2012**, *1*, 1–6. [CrossRef]

Disclaimer/Publisher's Note: The statements, opinions and data contained in all publications are solely those of the individual author(s) and contributor(s) and not of MDPI and/or the editor(s). MDPI and/or the editor(s) disclaim responsibility for any injury to people or property resulting from any ideas, methods, instructions or products referred to in the content.

Article

Effects on the Thermo-Mechanical and Interfacial Performance of Newly Developed PI-Sized Carbon Fiber–Polyether Ether Ketone Composites: Experiments and Molecular Dynamics Simulations

Hana Jung [†], Kwak Jin Bae [†], Yuna Oh, Jeong-Un Jin, Nam-Ho You and Jaesang Yu *

Composite Materials Application Research Center, Institute of Advanced Composite Materials, Korea Institute of Science and Technology, Chudong-ro 92, Bongdong-eup, Wanju-gun 55324, Republic of Korea
* Correspondence: jamesyu@kist.re.kr
† These authors contributed equally to this work.

Abstract: In this study, polyether ether ketone (PEEK) composites reinforced with newly developed water-dispersible polyimide (PI)-sized carbon fibers (CFs) were developed to enhance the effects of the interfacial interaction between PI-sized CFs and a PEEK polymer on their thermo-mechanical properties. The PI sizing layers on these CFs may be induced to interact vigorously with the p-phenylene groups of PEEK polymer chains because of increased electron affinity. Therefore, these PI-sized CFs are effective for improving the interfacial adhesion of PEEK composites. PEEK composites were reinforced with C-CFs, de-CFs, and PI-sized CFs. The PI-sized CFs were prepared by spin-coating a water-dispersible PAS suspension onto the de-CFs, followed by heat treatment for imidization. The composites were cured using a compression molding machine at a constant temperature and pressure. Atomic force and scanning electron microscopy observations of the structures and morphologies of the carbon fiber surfaces verified the improvement of their thermo-mechanical properties. Molecular dynamics simulations were used to investigate the effects of PI sizing agents on the stronger interfacial interaction energy between the PI-sized CFs and the PEEK polymer. These results suggest that optimal amounts of PI sizing agents increased the interfacial properties between the CFs and the PEEK polymer.

Keywords: carbon fiber; polyimide; sizing agent; interfacial adhesion; thermo-mechanical properties; molecular dynamics simulations

Citation: Jung, H.; Bae, K.J.; Oh, Y.; Jin, J.-U.; You, N.-H.; Yu, J. Effects on the Thermo-Mechanical and Interfacial Performance of Newly Developed PI-Sized Carbon Fiber–Polyether Ether Ketone Composites: Experiments and Molecular Dynamics Simulations. *Polymers* **2023**, *15*, 1646. https://doi.org/10.3390/polym15071646

Academic Editor: S. D. Jacob Muthu

Received: 27 February 2023
Revised: 21 March 2023
Accepted: 23 March 2023
Published: 25 March 2023

Copyright: © 2023 by the authors. Licensee MDPI, Basel, Switzerland. This article is an open access article distributed under the terms and conditions of the Creative Commons Attribution (CC BY) license (https://creativecommons.org/licenses/by/4.0/).

1. Introduction

In recent years, polyether ether ketone (PEEK) polymer composites reinforced with carbon fibers (CFs) have been widely used in high-performance applications, such as in the aerospace and automobile fields, due to their high heat resistance, very light weight, and great mechanical strength [1–14]. However, PEEK polymer has a high melting temperature of around 343 °C, which is constrained during the manufacturing process to improve the physical properties of the resulting composites [15–19]. Therefore, it was necessary to develop a sizing agent to protect the CFs with better thermal stability to increase the interfacial performance between the carbon fibers and matrix, which is an important factor in the load transfer of composites.

A variety of treatment methods (e.g., plasma, synthesis, and inducing functional groups of sizing agents on the fiber surfaces) have been studied [20–29]. Naito [30] demonstrated an increase in the tensile properties of composites reinforced with carbon nanotube (CNT)-grafted and polyimide (PI)-coated CFs. Hassan et al. [31] proposed that a loosely packed CNT network using PI could improve the interfacial adhesion between CFs and matrix due to enhanced CF wettability, polarity, and roughness. As a compatibilizer, PI exhibits a relatively high molecular weight in the interface regions as a result of a synergetic effect between PI and CNT. However, because PI is hard to dissolve, it requires a large amount of a strong organic solvent (such as N-Methyl-2-pyrrolidone) and limits the enhancement of interfacial adhesion [32–35]. Therefore, there is a need for novel research to create PI with high heat resistance for use as a sizing agent for carbon fibers.

Molecular dynamics (MD) simulations are attractive for predicting the effects of surface characteristics on the physical properties of polymer-based composites reinforced with carbon material [36–38]. It is possible to examine the thermal behavior of polymer-based composites by calculating the movement of molecules over time, which cannot be confirmed via experiments. Jiao et al. [39] and Jin et al. [40] studied the interfacial properties between polymers and sized-, functionalized-carbon materials using MD simulations. However, these results are not considered to prove enhanced interfacial properties without molecular mobility, which is significant in the thermal behavior of composites.

In this study, the thermo-mechanical properties and optimal sizing agent content of CF/PEEK composites with sizing agents, which can be applied even at the high processing temperature of PEEK polymers, were investigated. The aqueous PI sizing agent effects on the structure, morphology, interface performance, and thermo-mechanical properties of CF/PEEK composites were analyzed in detail to prove the comprehensive contributions of environmentally friendly modifications using various surface analysis techniques (Fourier-transform infrared spectroscopy (FT-IR), nuclear magnetic resonance (NMR), atomic force microscopy (AFM), and field-emission scanning electron microscopy (FE-SEM)) and physical estimates (interlaminar shear strength (ILSS), and dynamic mechanical analyzer (DMA) and thermogravimetric analyzers (TGA) characteristics). MD simulations were performed to analyze the interfacial characterization and thermal behavior according to the effect of the PI sizing agent between the CFs and the PEEK polymer. The calculated results can validate the experiments of the CF/PEEK composites.

2. Experiments

2.1. Materials

Pyromellitic dianhydride (PMDA, >99%), 4,4′-oxydianiline (ODA, >98%), and 4-hydroxy-1-methylpiperidine (HMP, >98%) were purchased from Tokyo Chemical Industry (Tokyo, Japan) to synthesize the poly (amic acid) salt (PAS). The N-Methyl-2-pyrrolidinone (NMP) was obtained from Sigma-Aldrich (St. Louis, MO, USA). A CF plain-weave fabric (C120-3K, plain, HD FIBER Co., Ltd., Yangsan, Republic of Korea) was used as reinforcement for the composites. C120-3K has a density of 1.78 g·cm^{-3}, a weight of 203 g·m^{-2}, and a thickness of 0.25 mm. The amorphous PEEK film (APTIV 2000, Victrex, Co., Ltd., Lancashire, UK) was used as a composite matrix with high heat resistance. The PEEK film has a density of 1.26 g·cm^{-3}, a tensile modulus of 1.80 GPa, a thickness of 16 μm, and a glass transition and melting temperature of 143 and 343 °C, respectively.

2.2. Preparation of Poly (Amic Acid) Salt (PAS) as a PI Sizing Agent

The polyimide (PI) was prepared from poly (amic acid) (PAA), a precursor, using a two-step reaction. The first step was the formation of the PAA solution via a nucleophilic substitution reaction of the monomer in a step-growth polymerization reaction. The second step was the dehydration and cyclization of the synthesized PAA solution with thermal treatment over 200 °C resulting in the final PI sizing agent. The synthetic process of the PAS powder was as follows (Figure 1). An amount of 0.15 mol of ODA was dissolved in 1200 mL of NMP at room temperature. Then, 0.15 mol of PMDA was added to the ODA solution, and the resulting mixture was stirred for 24 h under N$_2$ gas. This resulted in the formation of a PAA solution. Then, 0.3 mol of HMP was added to the PAA solution and stirred for 1 h. The mixture was then precipitated in acetone, and the resulting product was filtered, washed, and dried under vacuum at room temperature to obtain a PAS powder [15].

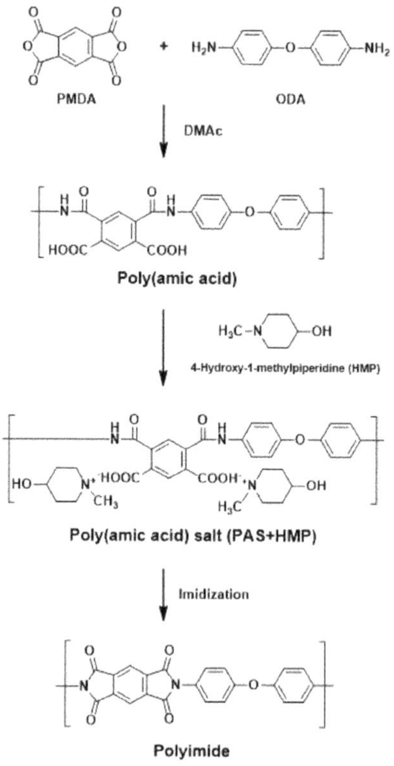

Figure 1. Synthetic route for PAS.

2.3. Fabrication of Water-Dispersible PI-Sized CF/PEEK Composites

Suspensions were created to deploy the sizing agent onto the carbon fiber surfaces using the PAS powder in distilled water. The water-dispersible PAS powder was heated in a water bath at 60 °C for 4 h to obtain the PI sizing agent, which was homogenized on a hot plate by stirring it with a magnetic bar. Sizing agents were prepared with PI contents of 0.5, 1.0, and 1.5 wt% compared with the CFs. The commercial carbon fabric (C-CF) was immersed in acetone at room temperature for 2 h to remove epoxy groups from the sizing agent. Then, the carbon fabric was cleaned in acetone in 2 steps (10 min each) and dried in a vacuum oven at 40 °C. The de-sized carbon fabric (de-CF) was modified by coating it with the prepared PAS suspensions using a spin coater at 800 rpm for 30 s. The PAS-modified CFs with different contents of 0.5, 1.0, and 1.5 wt% were dried in an oven at 40 °C for 12 h. For the imidization of the PAS-modified CF, heat treatment was carried out in an oven at 250 °C for 1 h. An illustration of the surface treatment of the CFs with the PI sizing agent is shown in Figure 2. In this study, the laminate was composed of 25 layers of carbon fiber fabric and 150 plies of PEEK film. PEEK composites were reinforced with C-CF, de-CF, and PI-sized CF. The laminate was preheated using a compression-molding machine at a constant temperature of 380 °C for 10 min without pressure. After the preheating process, the temperature was maintained at 380 °C with a pressure of 10 MPa for 20 min. The cooling step was performed using a water-cooling system with a rate of $-10\ °C \cdot min^{-1}$. The total weight fraction of the carbon fibers was about 58 wt%.

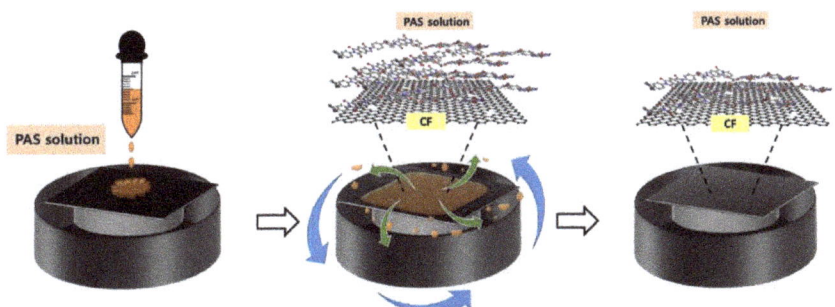

Figure 2. Schematic diagram of the process for surface treatment of carbon fibers using the PAS solution.

2.4. Characterizations

Nuclear magnetic resonance (NMR) spectra were determined with an Agilent 600 MHz Premium COMPACT spectrometer at 600 MHz for ^{1}H in dimethyl sulfoxide-d_6 (DMSO-d_6), using tetramethylsilane (TMS) as an internal standard. The attenuated total reflection–Fourier-transform infrared (ATR-FT-IR) spectra were obtained with a Nicolet IS10 with 32 scans per spectrum at a 2 cm^{-1} resolution. Atomic force microscopy (AFM) was conducted with a peak force tapping mode system (multimode-8, BRUKER Co., Billerica, MA, USA) to measure the surface roughness of the carbon fibers. Field-emission scanning electron microscopy (FE-SEM Verios 460L, FEI Corp., Hillsboro, OR, USA) with energy-dispersive X-ray spectroscopy (EDX) was performed to characterize the surface morphologies of the carbon fibers as well as the fracture surfaces of the composites reinforced with carbon fibers. The interlaminar shear strength (ILSS) of the composites was measured using the short-beam shear test as three-point bending according to the ASTM D 2344 standard (5569, Instron, Norwood, MA, USA) [41]. Specimens with the dimensions 28.2 mm × 9.4 mm × 4.7 mm were tested at a cross-head speed rate of 1 mm·min^{-1}. The short-beam strength (F^{sbs}) was calculated from the equation $F^{sbs} = 0.75\ P_b/bh$, where Pb represents the maximum load, b is the width, and h is the thickness of the specimen. The sample results were estimated as an average value from more than five specimens. The thermo-mechanical properties of the composites were obtained using a dynamic mechanical analyzer (DMA Q800, TA Instruments, New Castle, DE, USA) with a dual-cantilever mode at 1 Hz from 30 °C to 300 °C and a heating rate of 3 °C·min^{-1}, according to the ASTM 4065-01 standard [42]. The thermal stability of the carbon fibers was analyzed using thermogravimetric-differential thermal analysis (TG-DTA Analyzer SDT Q600, WATERS, Milford, MA, USA) with a temperature range of 25 °C to 800 °C with a heating rate of 10 °C min^{-1} under nitrogen gas.

2.5. Molecular Dynamics Modeling

MD simulations were performed to investigate the effect of the PI sizing agent on the interface between the CFs and PEEK polymer using Materials Studio 2017 software. The simulation model consisted of five graphene sheets, PI molecular chains, and PEEK molecular chains for modeling (Figure 3). The PI molecular chains with 0.0, 0.3, 1.0, 2.0, and 5.0 wt% compared with the carbon fibers were placed on the top surfaces of the graphene sheets. The simulation of the system based on a constant number, volume, and temperature (NVT) ensemble was performed to equilibrate all the models with a time step of 1.0 fs for 300 ps at 300 K. The simulation of the system based on a constant number, pressure, and temperature (NPT) ensemble was performed to further relax the internal stresses of the simulation models with an additional annealing process. All the models were heated from 300 K to 700 K at 50 K intervals and cooled down to 300 K at equal intervals. Each step was performed with a time step of 1.0 fs during 300 ps. The density of the equilibrated CF/PEEK models was 1.54 g·cm^{-3}, which was similar to the experimental result of 1.55 g·cm^{-3}. The dimension of the CF/PEEK composite model was 48.2 Å (x) × 49.9 Å (y) × 53.1 Å (z) in the periodic boundary condition. All the MD simulations were performed using the condensed-phase-optimized molecular potentials for the atomistic simulation studies (COMPASS II) force field [43], which has been widely used to describe the interaction between carbon materials and polymers.

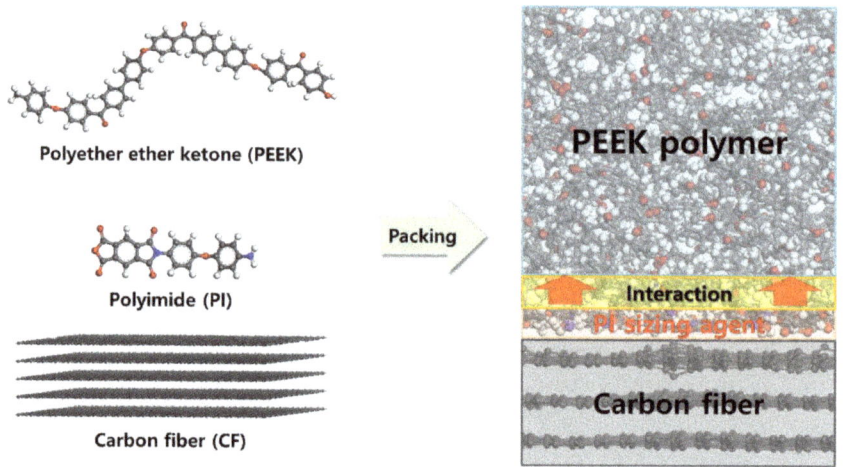

Figure 3. Molecular structures and composite models in the MD simulations.

The mean square displacement (MSD) can be used to characterize the speed at which particles move and quantify the mobility of molecules according to temperature. The MSD could be determined in the temperature range from 250 K to 500 K during the NPT ensemble. The MSD analysis can be calculated using the following equation:

$$\text{MSD} = \frac{1}{N}\sum_{i=1}^{N}\left|x^{(i)}(t) - x^{(i)}(0)\right|^2 \quad (1)$$

where N is the number of particles to be averaged, $x^{(i)}(0)$ is the reference position of the i-th particle, and $x^{(i)}(t)$ is the position of the i-th particle at time t.

2.6. Interfacial Characterization

The interaction energy of the CFs with the PEEK polymer was calculated using

$$\Delta E \ (kcal/mol) = E_{total} - \left(E_{CF} + E_{polymer}\right) \quad (2)$$

where ΔE is the interaction energy at the interface between the CFs and the PEEK polymer, E_{total} represents to the total potential energy of the composite, and E_{CF} and $E_{polymer}$ are the potential energy of the PI-sized CFs and PEEK polymer, respectively. The interfacial shear strength (ISS, τ) between the PI-sized CFs and the PEEK polymer was calculated using

$$\tau \ (MPa) = \frac{\Delta E}{WL^2} \quad (3)$$

where τ is the ISS, W is the width of the CF, L represents the length of the CF, and ΔE corresponds to the difference in the interaction energy.

3. Results and Discussion

3.1. Synthesis and Characterization of PAS

The ^{1}H NMR spectrum of PAS could be assigned by comparison between the ^{1}H NMR spectra of HMP and PAA, as shown in Figure 4. The marked numbers and letters in the Figure 4 were noticed each molecular group and spectrum peak. The peaks at 2.56 ppm in the ^{1}H NMR spectrum of HMP were related to the 3 protons of the CH_3 groups. The protons in PAA could be confirmed from the peaks at 6–9 ppm in the ^{1}H NMR spectrum of PAA. Those peaks were also found in the ^{1}H NMR spectrum of PAS. Moreover, the 6 peaks at 1.5–4 ppm in the ^{1}H NMR spectrum of PAS that appeared after the reaction of PAA with HMP were related to a proton on the piperidine ring of HMP. The proton of the CH_3 group in HMP was de-shielded by a proton of the COOH group in PAA.

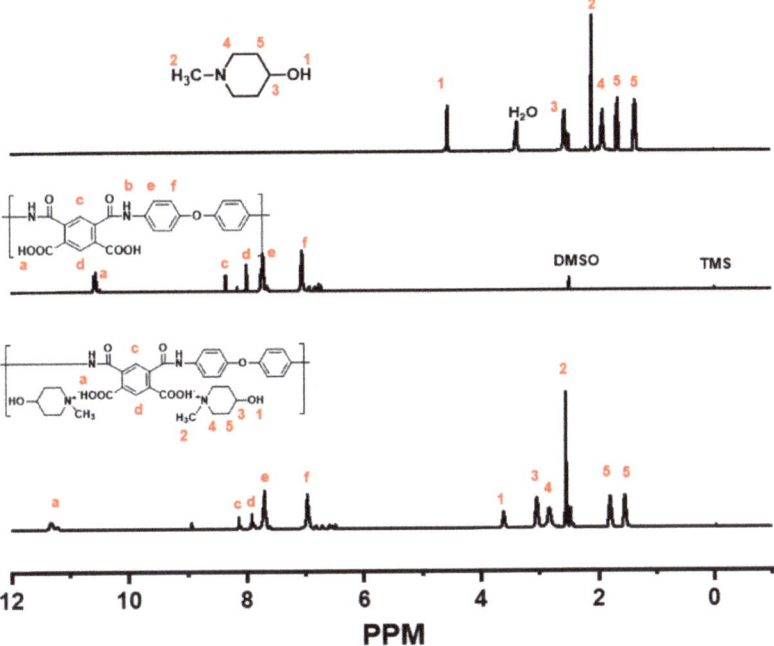

Figure 4. ^{1}H NMR spectra of HMP, PAA, and PAS.

Figure 5 shows the FT-IR spectra of PI, PAA, and PAS. The FT-IR spectrum of PAA shows N-H stretching at 3252 cm^{-1}, C=O stretching at 1620 cm^{-1}, N-H stretching at 1503 cm^{-1}, and C-O stretching at 1235 cm^{-1}. The FT-IR spectrum of PAS shows new bands at 3380 cm^{-1} and 2920 cm^{-1}, corresponding to the O-H and -CH_3 bonds of the HMP groups in PAS. The FT-IR spectrum of PAS, which can also be observed in the FT-IR spectrum of PAA, indicates that PAS was completely synthesized, as shown in Figure 5. The FT-IR spectra of PAA, PAS, and PI were used to verify the imidization of PAS to obtain cured PI. This was conducted via the observation of the disappearance of the peak at 1620 cm^{-1} and the growth of a peak at 1720 cm^{-1}. In Figure 5, the photographic image represents the aqueous PAS solution in water with an increase in concentration according to our previous study [15]. The synthesized PAS powder was readily available to the water-dispersible sizing agent without organic solvents.

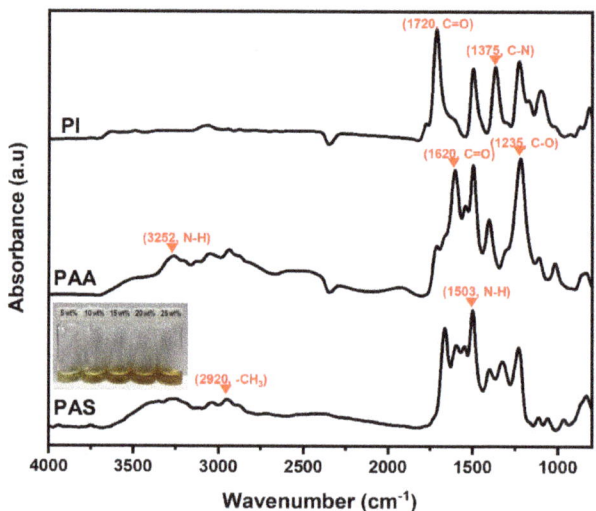

Figure 5. FT-IR spectra of PI, PAA, and PAS.

3.2. Surface Morphology Analysis of the PI-Sized CFs

The surface structures of the CFs were characterized using FT-IR, AFM, and FE-SEM observations. The FT-IR spectra of the C-CFs, de-CFs, and PI-sized CFs with contents of 0.5, 1.0, and 1.5 wt% were combined in Figure 6. The spectrum of the de-CFs was relatively smoother than that of the C-CFs due to the removal of epoxy groups and any other residues. A sharp carboxylic acid absorption peak (C=O stretching) in all the PI-sized CF samples was observed near 1700 cm^{-1}. The addition of PAS molecules onto the CF surfaces reveals N-H stretching. This was assigned to 1320 cm^{-1}, which was not displayed in the C-CFs. It could be determined that the carbon fiber surfaces were successfully modified with the cured PI from the newly synthesized water-dispersible PAS, which led to a more effective increase in the carbon fiber surface polarity and interfacial adhesion between the matrix and fibers.

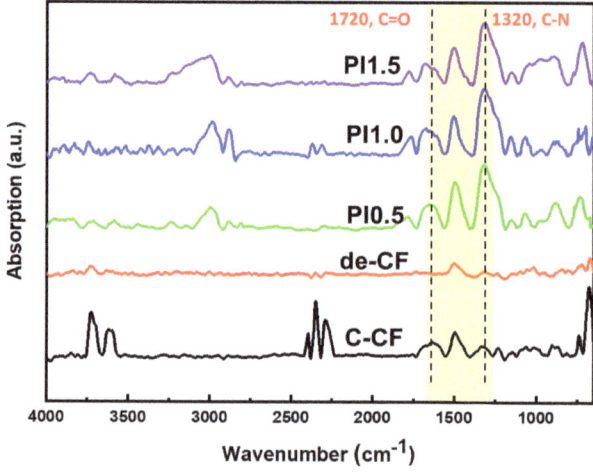

Figure 6. FT-IR spectra of C-CFs and de-CFs as well as 0.5, 1.0, and 1.5 wt% PI-sized CFs.

The 2D and 3D surface topographies of the de-CFs, C-CFs, and PI-sized CFs with different contents of PI sizing agents were compared, as shown in Figure 7. Increased surface roughness was observed on the PI-sized CF surfaces compared with those of the de-CFs. The average roughness (R_a) of the PI-sized CFs with contents of 0.5, 1.0, and 1.5 wt% (40.9, 62.1, and 44.0 nm, respectively) was greater than that of the de-CFs (32.5 nm). This increased roughness led to the enhancement of mechanical interlocking between the surface of the PI-sized CFs and the PEEK polymer [44]. However, in the case of the PI 1.5 wt% PI-sized CFs, the variation in height became relatively high (from −150 nm to 170 nm) due to an excessive amount of coating, as shown in Figure 7e [45,46]. This agglomerated non-uniform sizing layer between the carbon fibers and the matrix can lead to a decrease in interfacial adhesion.

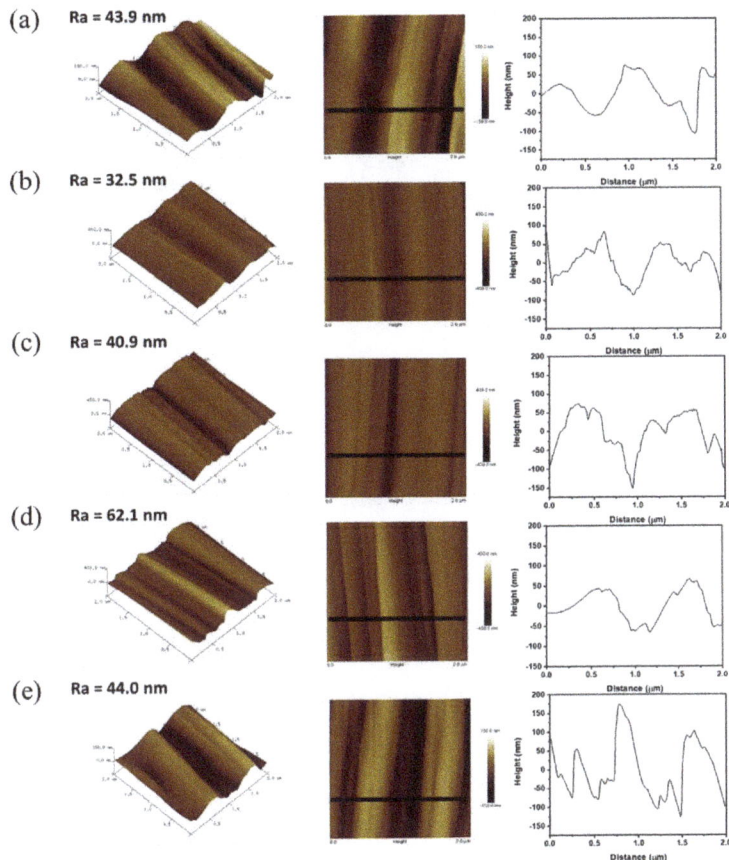

Figure 7. AFM images and height profiles for (**a**) C-CFs and (**b**) de-CFs, as well as (**c**) PI 0.5 wt%, (**d**) PI 1.0 wt%, and (**e**) PI 1.5 wt% of PI-sized CFs.

Observations using SEM-EDX were performed to analyze the compositional elements of reactive PI groups on the carbon fiber surfaces. Figure 8 shows the PI line profile of the EDX spectra across the carbon fiber surface along the scanline. The graph demonstrates the intensity of elemental carbon (C), nitrogen (N), and oxygen (O) with respect to distance along the fiber surface. The elements N and O had relatively uniform low concentrations over all of the carbon fiber surfaces. However, the C line profile shows an obvious decrease according to the increase in the content of the PI sizing agent, as shown in Figure 8a,e. This result was ascribed to a PI sizing layer containing uniformly attached N or O compared with the C-CFs or de-CFs. Increasing the amount of N or O greatly affects the interfacial adhesion between CFs and a PEEK polymer due to solubility with higher ionization [47].

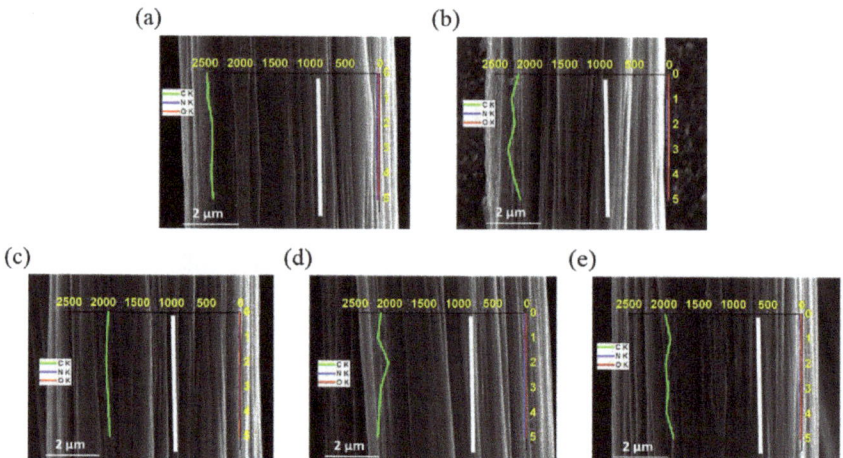

Figure 8. FE-SEM images and SEM-EDX line spectra of elemental carbon (green line), nitrogen (blue line), and oxygen (red line) in (**a**) C-CFs and (**b**) de-CFs, as well as (**c**) PI 0.5 wt%, (**d**) PI 1.0 wt%, and (**e**) PI 1.5 wt% PI-sized CFs.

3.3. Thermo-Mechanical Properties of the PI-Sized CF/PEEK Composites

The ILSS results were analyzed to investigate the characteristics of the interface between the CFs and the PEEK polymer, with or without the PI sizing agent, as shown in Figure 9. The short-beam strength for the PI 0.5-, 1.0-, and 1.5-sized CF-reinforced PEEK composites was 76.9, 73.8, and 71.4 MPa, respectively. The ILSSs of all PI-sized CFs and de-CF/PEEK composites were higher (up to 13.2%) than those of the C-CF/PEEK composites with the commercial sizing agents. The structure of the aromatic PI synthesized from the PAA salt of the PMDA/ODA type shows a charge-transfer interaction consisting of electron donors (including nitrogen groups) and electron acceptors (including carbonyl groups). This charge-transfer complex was attributed to the intramolecular forces between PI molecules, as well as to the intermolecular forces between PI molecules and PEEK polymer chains. These properties were supported by the strong molecular attraction due to the formation of robust polymer chains. With all these reactive groups, it was possible for them to attach layer by layer. Therefore, polyimides may not only form well-stacked sizing layers on the surfaces of carbon fibers but can also interact well with the p-phenylene groups of PEEK polymer chains [31]. This leads to an enhanced potential for the PI sizing agent to cause improved interfacial adhesion between the CFs and the PEEK polymer. The fiber surfaces of the de-CFs could not be protected from abrasion due to the lack of the sizing agent. In addition, the de-CFs were difficult to handle due to low cohesion. Therefore, the de-CFs are limited in fiber processing and production and composite material fabrication. Therefore, it is necessary for the treatment with a sizing agent to be appropriate to preserve the carbon fibers [48,49]. Figure 9 shows that the 0.5 wt% content of the PI sizing agent improved the short-beam strength of the CF/PEEK composites by approximately 13%. The commercial sizing agent is generally known to achieve approximately 1% coating on the surfaces of carbon fibers. This result demonstrates that composites reinforced with the newly developed water-dispersible PI have competitiveness for their efficient application to high-performance structures.

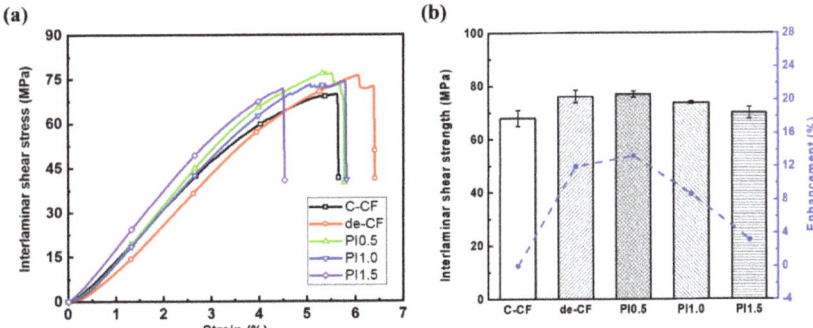

Figure 9. Interlaminar shear properties of composites reinforced with C-CFs and de-CFs, as well as PI 0.5 wt%, PI 1.0 wt%, and PI 1.5 wt% PI-sized CFs. (**a**) stress–strain curves; (**b**) interlaminar shear strengths and enhancements.

The DMA results on the viscoelastic properties of the composites reinforced with the C-CFs, de-CFs, and PI-sized CFs are shown in Figure 10. These results indicate that the complex modulus between the polymers can be affected by a wide variety of chain motions over a range of temperatures. Figure 10a shows a comparative increase in stored energy for all of the composites containing PI-sized CFs compared with the composites containing commercial epoxy-sized CFs. The storage moduli of the composites with 0.5, 1.0, and 1.5 wt% PI-sized CFs were 23,076, 24,440, and 23,724 MPa, respectively. In the case of the 1.0 wt% PI-sized CFs, the rate of increase in the storage modulus was approximately 11.6% compared with that of the C-CF/PEEK composites. As mentioned above, this is because the interfacial adhesion between the carbon fibers and the PEEK polymer was improved due to enhanced intermolecular attraction by adding the PI sizing agent. The effect of the PI sizing agent was demonstrated while ensuring excellent thermal stability at a high temperature. In addition, the increase in the loss modulus, which indicates energy dissipated as heat from the composites, approaches that of its viscous property. The composite reinforced with 1.0 wt% PI-sized CFs is distinctively represented in Figure 10b. This phenomenon means that the intermolecular force associated with the interfacial interaction between the PI sizing agent and the PEEK polymer was even greater than the intramolecular force between the CFs and the PI sizing agent of the PI-Sized CF/PEEK composites. This result indicates that the surface modification of the carbon fibers could be determined by the interfacial failure attributed to the physical properties of the composites. In addition, Figure 10c shows that the glass transition temperatures of the PI-Sized CF/PEEK composites were increased in comparison with the C-CF/PEEK composites. Among the content levels of the PI sizing agent, the appropriate amount on the surface of a CF is 0.5 wt% for the storage modulus as well as the damping factor (tan δ). This means that the thermal resistance to molecular mobility in the composites with the PI sizing agent was enhanced in comparison with the C-CFs or de-CF/PEEK composites [34]. Therefore, the effective incorporation of PI on the CFs was important to improve the thermo-mechanical properties of the PEEK composites reinforced with CFs. As shown in Figure 10c, the measured width from the tan δ curve for the composite reinforced C-CFs is remarkably wide, which indicates a greater distribution of the segment lengths related to molecular mobility than in the PI-Sized CF/PEEK composites. Moreover, the intensity of the tan δ curve for the composite-reinforced C-CFs presents a greater increase in the mobility of the relaxing segments of the polymer compared with that of the composite-reinforced PI-sized CFs. This refers to the weak interaction bonding resulting from poor inter–intramolecular forces between the C-CFs and the PEEK polymer with increasing temperature. These results suggest that the interfacial interaction between CFs and PEEK can be affected by the PI sizing agent on the surfaces of the carbon fibers. This effect was attributed to the significantly enhanced interfacial adhesion of the composites. The effect of the sizing agent with high heat resistance was the remarkable thermal stability of the CF/PEEK composites. In addition, the thermal stability of the PI-sized CFs with high heat resistance was compared with that of the C-CFs, as shown in Figure 11. The residual weights of the composites with C-CFs, de-CFs, PI0.5, PI1.0, and PI1.5 were 98.63, 99.55, 99.80, 99.59, and 99.67%, respectively. In the case of the PI-sized CFs with contents of 0.5, 1.0, and 1.5 wt%, there was no weight loss over the temperature range of 25 °C to 800 °C. The difference in the weight loss ratio between the PI-sized CFs and C-CFs was definitely observed depending on the absence of the polyimide sizing agent. Therefore, it was maintainable to protect the surfaces of the carbon fibers with the residual amount of the polyimide sizing agent after decomposition at a high temperature.

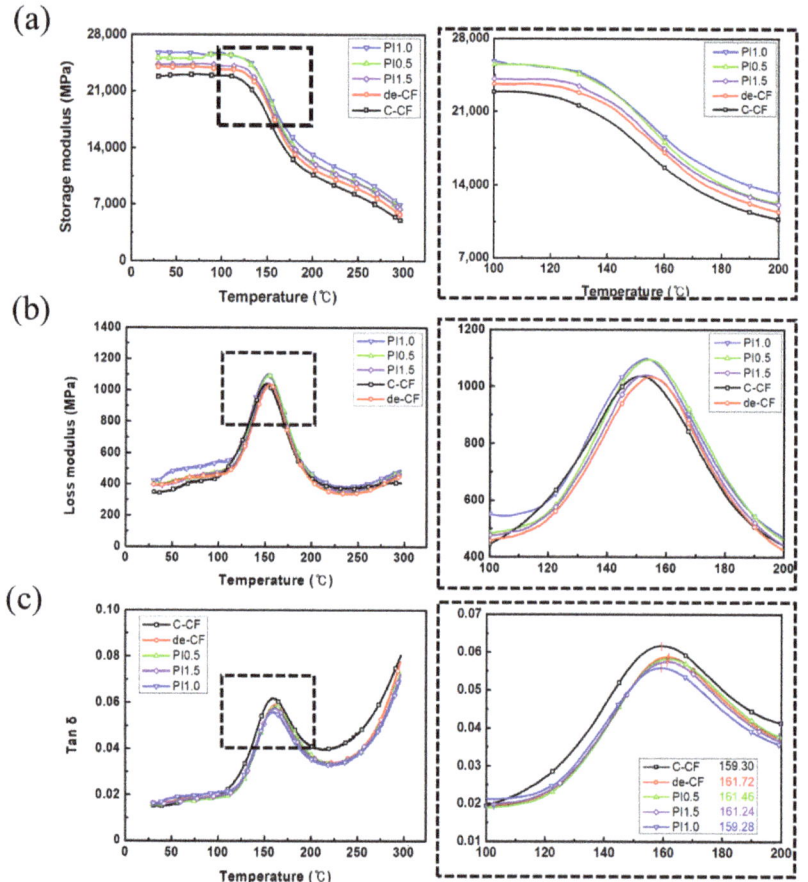

Figure 10. (a) Storage moduli, (b) loss moduli, and (c) tan δ curves for the composites reinforced with C-CFs and de-CFs, as well as PI 0.5 wt%, PI 1.0 wt%, and PI 1.5 wt% of PI-sized CFs.

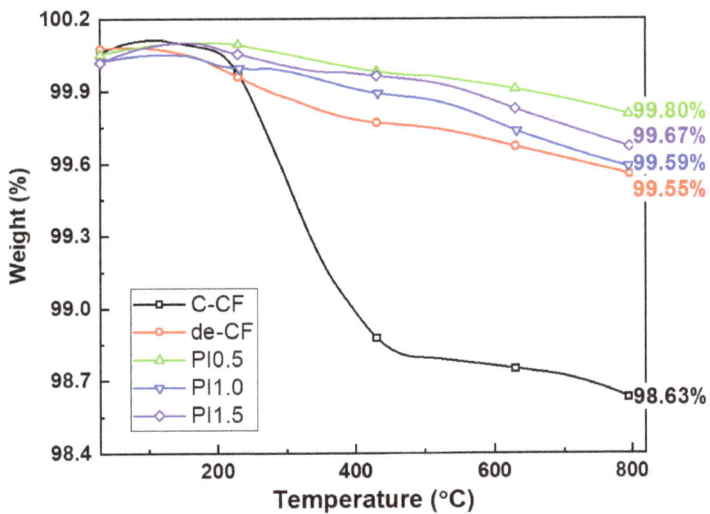

Figure 11. TGA curves of C-CFs, de-CFs, PI 0.5 wt%, PI 1.0 wt%, and PI 1.5 wt% sized CFs.

3.4. Fracture Characteristics of the CF/PEEK Composites

Figure 12 shows SEM images of the fracture surfaces on the composites reinforced with carbon fibers after the ILSS test. The fracture surfaces of the composites reinforced with C-CFs and de-CFs are relatively rough compared with those of the composites reinforced with PI-sized CFs at all content levels. Damaged fibers in the fill direction were rumpled, and pull-out of fibers were observed in the warp direction (Figure 12a,b). In addition, while delamination between the C-CFs or de-CFs and PEEK matrix was observed, the fracture surfaces of the PI 0.5, 1.0, and 1.5 wt% PEEK composites are rather smooth and tightly embedded between the CFs and the PEEK matrix. This phenomenon confirms that the interfacial adhesion was enhanced by adding the PI sizing agent to the carbon fibers [50]. The PI sizing agent was protected due to its high heat resistance during high-temperature manufacturing processes, which allowed it to withstand the effective stress transfer of the carbon fiber. The PI-sized CFs played an important role in improving the interfacial performance of these PEEK composites.

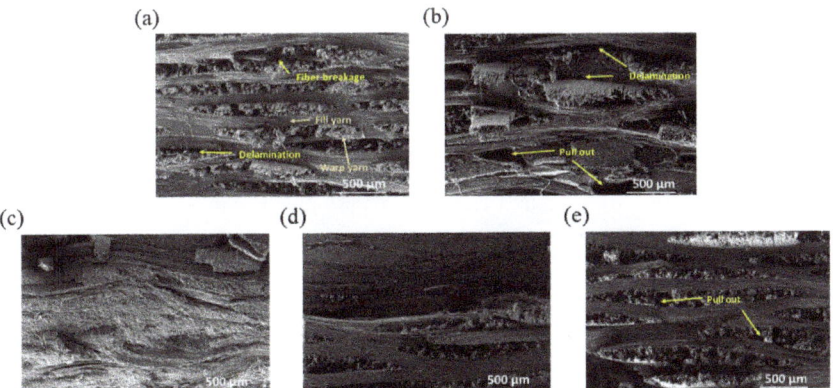

Figure 12. FE-SEM images of fracture surfaces on composites reinforced with (**a**) C-CFs and (**b**) de-CFs, as well as (**c**) PI 0.5 wt%, (**d**) PI 1.0 wt%, and (**e**) PI 1.5 wt% of the PI-sized CFs.

3.5. Molecular Dynamics of the CF/PEEK Composite

MSD analysis can derive an estimation of the parameters of molecular movements, such as the diffusion coefficient. When the model systems reached the phase transition region, the diffusion level increased sharply, which can determine the glass transition temperature. The temperatures of the phase transition regions of the composite models were increased, and are marked with a red arrow in Figure 13. The temperatures of the phase transition regions of the composite models with de-CFs, PI0.3, and PI1.0 were 400–450 K, 425–475 K, and 450–475 K, respectively. The de-CFs (Figure 13a) show a relatively low temperature range compared with the PI-sized composites. This result shows that the PI sizing agent increased the glass transition temperature.

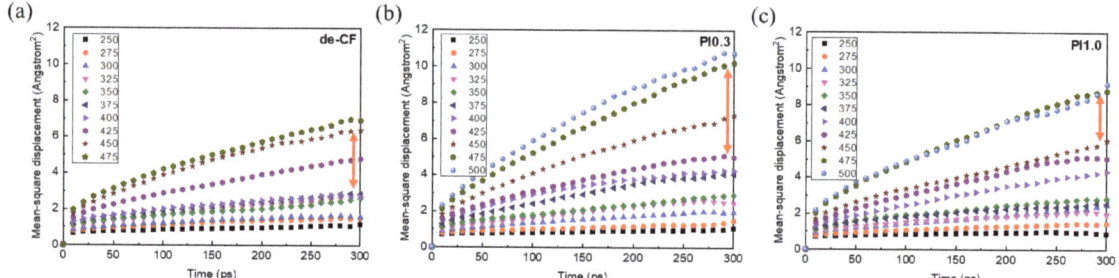

Figure 13. Mean square displacement (MSD) results: (**a**) de-CFs, (**b**) PI0.3, and (**c**) PI1.0.

The interfacial characteristics between the de-CFs and the PEEK polymer were calculated using MD simulations. Table 1 shows the interaction energy (ΔE) between the CFs and the PEEK polymer. The negative interfacial energy means that the energy of adsorption of the PEEK molecules at the interface was strong [51,52]. Therefore, the high negative value indicates strong adhesion at the interface between the CFs and the PEEK polymer. The interaction energy of the de-CF/PEEK composites was −1788.79 kcal/mol. Those of the PI-Sized CF/PEEK composites with contents of PI 0.3, 1.0, 2.0, and 5.0 wt% were −1877.73, −1869.48, −1815.94, and −1767.29 kcal/mol, respectively. These results show that the interaction energies of the PI-Sized CF/PEEK composites with contents of PI 0.3, 1.0, and 2.0 wt% generally had a higher interaction energy than that of the de-CF/PEEK composite. In addition, the PI sizing agents also affected the interfacial shear strength (ISS), as shown in Figure 14. The interfacial shear strengths of the de-CFs, PI0.3, PI1.0, PI2.0, and PI5.0 were calculated to be 103.15, 107.17, 107.31, 103.84, and 101.24 MPa, respectively, obtained from Equation (3). The simulated interfacial shear strengths of the PI-Sized CF/PEEK composites, except for the PI5.0 model, were higher than that of the de-CF/PEEK composite. This result proves that the PI sizing agent affected the interfacial properties between the CFs and the PEEK polymer. In addition, the contents of 0.3 wt% and 1.0 wt% of the PI sizing agent on the CFs were the most effective for improving the interaction energy and interfacial shear strength. The results obtained from the MD simulations show similar trends to the experimental data.

Table 1. Interaction energy of CF/PEEK composites using MD simulation.

Sample (300 K)		E_{total} (kcal/mol)	E_{CF} (kcal/mol)	$E_{polymer}$ (kcal/mol)	ΔE (kcal/mol)
de-CF/PEEK		243,256.35	229,233.60	15,811.53	−1788.79
PI-Sized CF/PEEK	0.3 wt%	243,991.70	229,315.41	16,554.02	−1877.73
	1.0 wt%	244,734.37	229,173.83	17,430.01	−1869.48
	2.0 wt%	242,697.81	229,127.90	15,385.85	−1815.94
	5.0 wt%	240,873.99	228,557.79	14,083.48	−1767.29

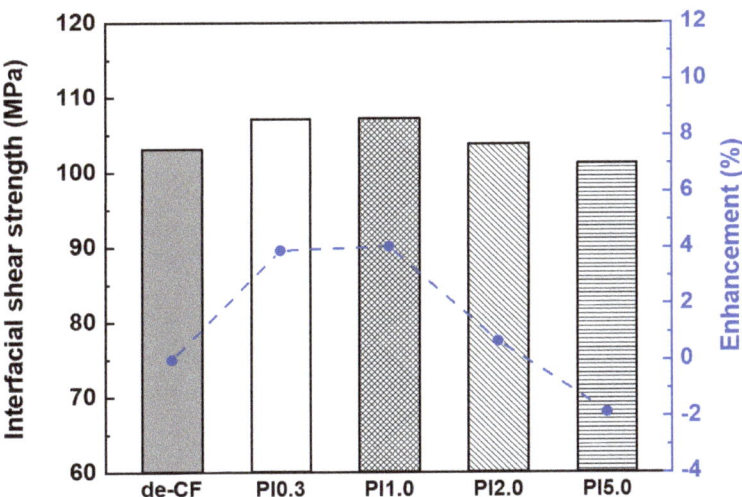

Figure 14. Interfacial shear strength of PEEK composites reinforced with de-CFs as well as PI 0.3 wt%, PI 1.0 wt%, PI 2.0 wt%, and PI 5.0 wt% sized CFs using MD simulation.

4. Conclusions

In this study, the effects of water-dispersible PI-sized CFs on the thermo-mechanical properties of PEEK composites were investigated. The PMDA/ODA used for the PAA solution was successfully synthesized and thermally imidized to apply it as a sizing agent for the carbon fibers. The surfaces of the carbon fibers were modified by adding a PI sizing layer that led to improved interfacial adhesion between the CFs and PEEK matrix. The short-beam strengths and storage moduli of the PEEK composites reinforced with PI-sized CFs increased more than those of the composites reinforced with de-sized and commercial CFs. The surface roughness of the PI-sized CFs as indicated in the 2D and 3D topographies was greater than those of the C-CFs and de-CFs. This result shows that the CF surfaces were coated with the PI sizing agent. The fracture surfaces of the composites reinforced with PI-sized CFs exhibited greater smoothness and were more tightly embedded between the CFs and PEEK matrix than those of the composites reinforced with C-CFs and de-CFs. This was ascribed to enhanced mechanical interlocking between the PI-sized CFs and the PEEK polymer of the composites. These results suggest that the interfacial interaction between CFs and PEEK can be affected by a PI sizing agent on the surfaces of carbon fibers. This effect was attributed to the significantly enhanced interfacial adhesion of the composites even at a high temperature. MD simulations were performed to examine the effects of the PI sizing agents on the interfacial properties between the CFs and the PEEK polymer. The interfacial properties between the CFs and the PEEK polymer were most improved when the PI sizing agent with 0.3 wt% was applied to the surface of the carbon fibers. As a result, the newly developed PI sizing agent with high heat resistance can be used to improve the interfacial bonding properties and thermo-mechanical properties of CF/PEEK composites. The present study focused on improving the interfacial bonding properties and thermo-mechanical properties of CF/PEEK composites. PI-sized carbon nanomaterials with high-temperature resistance and high dispersion can be investigated to improve the thermo-mechanical properties of polymer-based composites in the future.

Author Contributions: Conceptualization, H.J. and K.J.B.; methodology, H.J. and J.-U.J.; software, K.J.B. and Y.O.; validation, N.-H.Y. and J.Y.; formal analysis, K.J.B. and H.J.; investigation, K.J.B.; data curation, H.J.; writing-original draft preparation, H.J. and K.J.B.; writing-review and editing, N.-H.Y. and J.Y.; supervision, J.Y.; project administration, J.Y.; funding acquisition, J.Y., H.J. and K.J.B. All authors contributed equally to this paper. All authors have read and agreed to the published version of the manuscript.

Funding: This study was supported by the Korea Institute of Science and Technology (KIST) Institutional Program. This study was also supported by the Technology Innovation Program 'Development of an ultra-lightweight 19 carbon composite wheel using integrated braid preform manufacturing technology with tow prepreg material' (no. 20021913) funded by the Ministry of Trade, Industry, and Energy (MOTIE, Korea).

Institutional Review Board Statement: Not applicable.

Informed Consent Statement: Not applicable.

Data Availability Statement: The data presented in this study are available on request from the corresponding author.

Acknowledgments: This study was supported by the Korea Institute of Science and Technology (KIST) Institutional Program.

Conflicts of Interest: The authors declare no conflict of interest.

References

1. Wang, X.K.; Huang, Z.G.; Lai, M.L.; Jiang, L.; Zhang, Y.; Zhou, H.M. Highly enhancing the interfacial strength of CF/PEEK composites by introducing PAIK onto diazonium functionalized carbon fibers. *Appl. Surf. Sci.* **2020**, *510*, 145400. [CrossRef]
2. Yan, T.W.; Yan, F.; Li, S.Y.; Li, M.; Liu, Y.; Zhang, M.J.; Jin, L.; Shang, L.; Liu, L.; Ao, Y.H. Interfacial enhancement of CF/PEEK composites by modifying water-based PEEK-NH2 sizing agent. *Compos. B Eng.* **2020**, *199*, 108258. [CrossRef]
3. Stepashkin, A.A.; Chukov, D.I.; Senatov, F.S.; Salimon, A.I.; Korsunsky, A.M.; Kaloshkin, S.D. 3D-printed PEEK-carbon fiber (CF) composites: Structure and thermal properties. *Compos. Sci. Technol.* **2018**, *164*, 319–326. [CrossRef]
4. Yan, M.X.; Tian, X.Y.; Peng, G.; Li, D.C.; Zhang, X.Y. High temperature rheological behavior and sintering kinetics of CF/PEEK composites during selective laser sintering. *Compos. Sci. Technol.* **2018**, *165*, 140–147. [CrossRef]
5. Pan, L.; Yapici, U. A comparative study on mechanical properties of carbon fiber/PEEK composites. *Adv. Compos. Mater.* **2016**, *25*, 359–374. [CrossRef]
6. Sharma, M.; Bijwe, J.; Mader, E.; Kunze, K. Strengthening of CF/PEEK interface to improve the tribological performance in low amplitude oscillating wear mode. *Wear* **2013**, *301*, 735–739. [CrossRef]
7. Molazemhosseini, A.; Tourani, H.; Khavandi, A.; Yekta, B.E. Tribological performance of PEEK based hybrid composites reinforced with short carbon fibers and nano-silica. *Wear* **2013**, *303*, 397–404. [CrossRef]
8. Zhao, W.Y.; Yu, R.; Dong, W.Y.; Luan, J.S.; Wang, G.B.; Zhang, H.B.; Zhang, M. The influence of long carbon fiber and its orientation on the properties of three-dimensional needle-punched CF/PEEK composites. *Compos. Sci. Technol.* **2021**, *203*, 108565. [CrossRef]
9. Lessard, H.; Lebrun, G.; Benkaddour, A.; Pham, X.T. Influence of process parameters on the thermo-stamping of a [0/90]$_{(12)}$ carbon/polyether ether ketone laminate. *Compos. Part. A Appl. Sci. Manuf.* **2015**, *70*, 59–68. [CrossRef]
10. Vieille, B.; Albouy, W.; Taleb, L. Influence of stamping on the compressive behavior and the damage mechanisms of C/PEEK laminates bolted joints under severe conditions. *Compos. B Eng.* **2015**, *79*, 631–638. [CrossRef]
11. Stokes-Griffin, C.M.; Compston, P. The effect of processing temperature and placement rate on the short beam strength of carbon fibre-PEEK manufactured using a laser tape placement process. *Compos. Part. A Appl. Sci. Manuf.* **2015**, *78*, 274–283. [CrossRef]
12. Dworak, M.; Rudawski, A.; Markowski, J.; Blazewicz, S. Dynamic mechanical properties of carbon fibre-reinforced PEEK composites in simulated body-fluid. *Compos. Struct.* **2017**, *161*, 428–434. [CrossRef]
13. Na, R.Q.; Liu, J.Y.; Wang, G.B.; Zhang, S.L. Light weight and flexible poly (ether ether ketone) based composite film with excellent thermal stability and mechanical properties for wide-band electromagnetic interference shielding. *RSC Adv.* **2018**, *8*, 3296–3303. [CrossRef] [PubMed]
14. Yang, Y.C.; Wang, T.J.; Wang, S.D.; Cong, X.; Zhang, S.L.; Zhang, M.; Luan, J.S.; Wang, G.B. Strong Interface Construction of Carbon Fiber-reinforced PEEK Composites: An Efficient Method for Modifying Carbon Fiber with Crystalline PEEK. *Macromol. Rapid Commun.* **2020**, *41*, 2000001. [CrossRef]
15. Jung, H.; Bae, K.J.; Jin, J.; Oh, Y.; Hong, H.; You, N.H.; Yu, S.J. The effect of aqueous polyimide sizing agent on PEEK based carbon fiber composites using experimental techniques and molecular dynamics simulations. *Func. Compos. Struct.* **2020**, *2*, 025001. [CrossRef]

16. Gao, X.P.; Huang, Z.G.; Zhou, H.M.; Li, D.Q.; Li, Y.; Wang, Y.M. Higher mechanical performances of CF/PEEK composite laminates via reducing interlayer porosity based on the affinity of functional s-PEEK. *Polym. Compos.* **2019**, *40*, 3749–3757. [CrossRef]
17. Lu, C.R.; Xu, N.; Zheng, T.; Zhang, X.; Lv, H.X.; Lu, X.; Xiao, L.; Zhang, D.X. The Optimization of Process Parameters and Characterization of High-Performance CF/PEEK Composites Prepared by Flexible CF/PEEK Plain Weave Fabrics. *Polymers* **2019**, *11*, 53. [CrossRef]
18. Chang, B.N.; Li, X.M.; Parandoush, P.; Ruan, S.L.; Shen, C.Y.; Lin, D. Additive manufacturing of continuous carbon fiber reinforced poly-ether-ether-ketone with ultrahigh mechanical properties. *Polym. Test.* **2020**, *88*, 106563. [CrossRef]
19. Yao, S.S.; Jin, F.L.; Rhee, K.Y.; Hui, D.; Park, S.J. Recent advances in carbon-fiber-reinforced thermoplastic composites: A review. *Compos. B Eng.* **2018**, *142*, 241–250. [CrossRef]
20. Hassan, E.A.M.; Yang, L.L.; Elagib, T.H.H.; Ge, D.T.; Lv, X.W.; Zhou, J.F.; Yu, M.H.; Zhu, S. Synergistic effect of hydrogen bonding and pi-pi stacking in interface of CF/PEEK composites. *Compos. B Eng.* **2019**, *171*, 70–77. [CrossRef]
21. Guo, L.H.; Zhang, G.; Wang, D.A.; Zhao, F.Y.; Wang, T.M.; Wang, Q.H. Significance of combined functional nanoparticles for enhancing tribological performance of PEEK reinforced with carbon fibers. *Compos. Part. A Appl. Sci. Manuf.* **2017**, *102*, 400–413. [CrossRef]
22. Zhang, G. Structure-Tribological Property Relationship of Nanoparticles and Short Carbon Fibers Reinforced PEEK Hybrid Composites. *J. Polym. Sci. B Pol. Phys.* **2010**, *48*, 801–811. [CrossRef]
23. Liu, H.S.; Zhao, Y.; Li, N.; Zhao, X.R.; Han, X.; Li, S.; Lu, W.K.; Wang, K.; Du, S.Y. Enhanced interfacial strength of carbon fiber/PEEK composites using a facile approach via PEI&ZIF-67 synergistic modification. *J. Mater. Res. Technol.* **2019**, *8*, 6289–6300. [CrossRef]
24. Chen, J.L.; Wang, K.; Zhao, Y. Enhanced interfacial interactions of carbon fiber reinforced PEEK composites by regulating PEI and graphene oxide complex sizing at the interface. *Compos. Sci. Technol.* **2018**, *154*, 175–186. [CrossRef]
25. Moosburger-Will, J.; Bauer, M.; Laukmanis, E.; Horny, R.; Wetjen, D.; Manske, T.; Schmidt-Stein, F.; Topker, J.; Horn, S. Interaction between carbon fibers and polymer sizing: Influence of fiber surface chemistry and sizing reactivity. *Appl. Surf. Sci.* **2018**, *439*, 305–312. [CrossRef]
26. Jang, J.; Kim, H. Improvement of carbon fiber/PEEK hybrid fabric composites using plasma treatment. *Polym. Compos.* **1997**, *18*, 125–132. [CrossRef]
27. Mao, J.H.; Pan, Y.S.; Ding, J. Tensile mechanical characteristics of CF/PEEK biocomposites with different surface modifications. *Micro Nano Lett.* **2019**, *14*, 263–268. [CrossRef]
28. Zabihi, O.; Ahmadi, M.; Li, Q.; Shafei, S.; Huson, M.G.; Naebe, M. Carbon fibre surface modification using functionalized nanoclay: A hierarchical interphase for fibre-reinforced polymer composites. *Compos. Sci. Technol.* **2017**, *148*, 49–58. [CrossRef]
29. Li, Q.X.; Church, J.S.; Naebe, M.; Fox, B.L. Interfacial characterization and reinforcing mechanism of novel carbon nanotube-Carbon fibre hybrid composites. *Carbon* **2016**, *109*, 74–86. [CrossRef]
30. Naito, K. Tensile Properties of Polyimide Composites Incorporating Carbon Nanotubes-Grafted and Polyimide-Coated Carbon Fibers. *J. Mater. Eng. Perform.* **2014**, *23*, 3245–3256. [CrossRef]
31. Hassan, E.A.M.; Ge, D.T.; Zhu, S.; Yang, L.L.; Zhou, J.F.; Yu, M.H. Enhancing CF/PEEK composites by CF decoration with polyimide and loosely-packed CNT arrays. *Compos. Part. A Appl. Sci. Manuf.* **2019**, *127*, 105613. [CrossRef]
32. Kim, H.; Ku, B.C.; Goh, M.; Ko, H.C.; Ando, S.; You, N.H. Synergistic Effect of Sulfur and Chalcogen Atoms on the Enhanced Refractive Indices of Polyimides in the Visible and Near-Infrared Regions. *Macromolecules* **2019**, *52*, 827–834. [CrossRef]
33. He, S.Q.; Zhang, S.C.; Lu, C.X.; Wu, G.P.; Yang, Y.; An, F.; Guo, J.H.; Li, H. Polyimide nano-coating on carbon fibers by electrophoretic deposition. *Colloids Surf. A Physicochem. Eng. Asp.* **2011**, *381*, 118–122. [CrossRef]
34. Wang, T.; Jiao, Y.S.; Mi, Z.M.; Li, J.T.; Wang, D.R.N.; Zhao, X.G.; Zhou, H.W.; Chen, C.H. PEEK composites with polyimide sizing SCF as reinforcement: Preparation, characterization, and mechanical properties. *High Perform. Polym.* **2020**, *32*, 383–393. [CrossRef]
35. Naganuma, T.; Naito, K.; Yang, J.M. High-temperature vapor deposition polymerization polyimide coating for elimination of surface nano-flaws in high-strength carbon fiber. *Carbon* **2011**, *49*, 3881–3890. [CrossRef]
36. Jung, H.; Choi, H.K.; Oh, Y.; Hong, H.; Yu, J. Enhancement of thermomechanical stability for nanocomposites containing plasma treated carbon nanotubes with an experimental study and molecular dynamics simulations. *Sci. Rep.* **2020**, *10*, 405. [CrossRef]
37. Jung, H.; Choi, H.K.; Kim, S.; Lee, H.S.; Kim, Y.; Yu, J. The influence of N-doping types for carbon nanotube reinforced epoxy composites: A combined experimental study and molecular dynamics simulation. *Compos. Part A Appl. Sci. Manuf. S.* **2017**, *103*, 17–24. [CrossRef]
38. Li, Y.L.; Wang, S.J.; Wang, Q. A molecular dynamics simulation study on enhancement of mechanical and tribological properties of polymer composites by introduction of graphene. *Carbon* **2017**, *111*, 538–545. [CrossRef]
39. Jiao, W.W.; Hou, C.L.; Zhang, X.H.; Liu, W.B. Molecular dynamics simulation of the influence of sizing agent on the interfacial properties of sized carbon fiber/vinyl ester resin composite modified by self-migration method. *Compos. Interfaces* **2021**, *28*, 445–459. [CrossRef]
40. Jin, Y.K.; Duan, F.L.; Mu, X.J. Functionalization enhancement on interfacial shear strength between graphene and polyethylene. *Appl. Surf. Sci.* **2016**, *387*, 1100–1109. [CrossRef]

41. *ASTM D2344/D2344M*; Standard Test Method for Short-Beam Strength of Polymer Matrix Composite Materials and Their Laminates; Annual Book of ASTM Standards. ASTM: West Conshohocken, PA, USA, 2013.
42. *ASTM D4065*; Standard Practice for Plastics: Dynamic Mechanical Properties: Determination and Report of Procedures; Annual Book of ASTM Standards. ASTM: West Conshohocken, PA, USA, 2001.
43. Sun, H. COMPASS: An ab initio force-field optimized for condensed-phase applications-Overview with details on alkane and benzene compounds. *J. Phys. Chem. B* **1998**, *102*, 7338–7364. [CrossRef]
44. Yao, Y.; Chen, S.H. The effects of fiber's surface roughness on the mechanical properties of fiber-reinforced polymer composites. *J. Compos. Mater.* **2013**, *47*, 2909–2923. [CrossRef]
45. Liu, H.S.; Zhao, Y.; Li, N.; Li, S.; Li, X.K.; Liu, Z.W.; Cheng, S.; Wang, K.; Du, S.Y. Effect of polyetherimide sizing on surface properties of carbon fiber and interfacial strength of carbon fiber/polyetheretherketone composites. *Polym. Compos.* **2021**, *42*, 931–943. [CrossRef]
46. Kong, D.C.; Yang, M.H.; Zhang, X.S.; Du, Z.C.; Fu, Q.; Gao, X.Q.; Gong, J.W. Control of Polymer Properties by Entanglement: A Review. *Macromol. Mater. Eng.* **2021**, *306*, 2100536. [CrossRef]
47. Raghavendran, V.K.; Drzal, L.T.; Askeland, P. Effect of surface oxygen content and roughness on interfacial adhesion in carbon fiber-polycarbonate composites. *J. Adhes. Sci. Technol.* **2002**, *16*, 1283–1306. [CrossRef]
48. Zheng, H.; Zhang, W.J.; Li, B.W.; Zhu, J.J.; Wang, C.H.; Song, G.J.; Wu, G.S.; Yang, X.P.; Huang, Y.D.; Ma, L.C. Recent advances of interphases in carbon fiber-reinforced polymer composites: A review. *Compos. B Eng.* **2022**, *233*, 109639. [CrossRef]
49. Tang, L.G.; Kardos, J.L. A review of methods for improving the interfacial adhesion between carbon fiber and polymer matrix. *Polym. Compos.* **1997**, *18*, 100–113. [CrossRef]
50. Hwang, D.; Lee, S.G.; Cho, D. Dual-Sizing Effects of Carbon Fiber on the Thermal, Mechanical, and Impact Properties of Carbon Fiber/ABS Composites. *Polymers* **2021**, *13*, 2298. [CrossRef]
51. Sun, S.Q.; Chen, S.H.; Weng, X.Z.; Shan, F.; Hu, S.Q. Effect of Carbon Nanotube Addition on the Interfacial Adhesion between Graphene and Epoxy: A Molecular Dynamics Simulation. *Polymers* **2019**, *11*, 121. [CrossRef]
52. Zuo, Z.; Liang, L.F.; Bao, Q.Q.; Yan, P.T.; Jin, X.; Yang, Y.L. Molecular Dynamics Calculation on the Adhesive Interaction Between the Polytetrafluoroethylene Transfer Film and Iron Surface. *Front. Chem.* **2021**, *9*, 740447. [CrossRef]

Disclaimer/Publisher's Note: The statements, opinions and data contained in all publications are solely those of the individual author(s) and contributor(s) and not of MDPI and/or the editor(s). MDPI and/or the editor(s) disclaim responsibility for any injury to people or property resulting from any ideas, methods, instructions or products referred to in the content.

Article

Effect of Short Fibers on Fracture Properties of Epoxy-Based Polymer Concrete Exposed to High Temperatures

Oussama Elalaoui [1,2]

1. Department of Civil and Environmental Engineering, College of Engineering, Majmaah University, Al-Majmaah 11952, Saudi Arabia; o.elalaoui@mu.edu.sa or elalaoui_o@yahoo.fr
2. Civil Engineering Laboratory, National Engineering School of Tunis, University of El Manar, BP 37, Tunis 1002, Tunisia

Abstract: Recently, polymer concrete (PC) has been widely used in many civil engineering applications. PC shows superiority in major physical, mechanical, and fracture properties comparing to ordinary Portland cement concrete. Despite many suitable characteristics of thermosetting resins related to processing, the thermal resistance of polymer concrete composite is relatively low. This study aims to investigate the effect of incorporating short fibers on mechanical and fracture properties of PC under different ranges of high temperatures. Short carbon and polypropylene fibers were added randomly at a rate of 1 and 2% by the total weight of the PC composite. The exposure temperatures cycles were ranged between 23 to 250 °C. Various tests were conducted including flexure strength, elastic modulus, toughness, tensile crack opening, density, and porosity to evaluate the effect of addition of short fibers on fracture properties of PC. The results show that the inclusion of short fiber lead to an increase in the load carrying capacity of PC by an average of 24% and limits the crack propagation. On the other hand, the enhancement of fracture properties of based PC containing short fibers is vanished at high temperature (250 °C), but still more efficient than ordinary cement concrete. This work could lead to broader applications of polymer concrete exposed to high temperatures.

Keywords: polymer concrete; short fibers; fracture properties; thermal resistance; porosity

Citation: Elalaoui, O. Effect of Short Fibers on Fracture Properties of Epoxy-Based Polymer Concrete Exposed to High Temperatures. *Polymers* **2023**, *15*, 1078. https://doi.org/10.3390/polym15051078

Academic Editors: S. D. Jacob Muthu and Victor Tcherdyntsev

Received: 29 December 2022
Revised: 25 January 2023
Accepted: 2 February 2023
Published: 21 February 2023

Copyright: © 2023 by the author. Licensee MDPI, Basel, Switzerland. This article is an open access article distributed under the terms and conditions of the Creative Commons Attribution (CC BY) license (https://creativecommons.org/licenses/by/4.0/).

1. Introduction

Polymer concrete (PC), one of the three types of polymer concrete type, is formed by completely substituting Portland cement with polymer binders. The two other types are polymer-impregnated concrete (PIC), which consists of a hardened Portland cement concrete impregnated with a low-viscosity monomer polymerized in situ, and the second one is the polymer cement concrete (PCC) which is produced by incorporating a monomer or polymer in a Portland cement concrete mix and polymerized after placing concrete [1].

The polymer binder used in PC is acting through a polymerization process to bond the solid particles of concrete [2,3]. PC, whether reinforced or non-reinforced, shows superiority in major physical, mechanical, and fracture properties when compared to ordinary Portland cement. These properties include high mechanical strength, rapid hardening, and short curing time, which are very beneficial for precast element production or partitions with high toughness. [4]. In addition, PC is attributed with fast demolding, minimum cracking, higher mechanical strength, good bonding to old substrates, better abrasion resistance, chemical resistance, and less maintenance in comparison to Portland cement concrete [5,6]. This explains the quick development of using PC in many civil engineering applications [7].

Various types of resins have been used in the production of polymer concrete such as epoxy [8,9], unsaturated polyester [10,11], acrylics and vinyl-ester, poly acrylate, polypropylene, furan-based resin, and other polymers [12,13]. However, the formed composites incorporating these materials are brittle and the majority of their properties are very sensitive to long exposure at high temperatures [14–17]. This is because the polymers acting as binders in the PC are organic substances, which exhibits lower resistant to heat than the inorganic ones. Accordingly, PC is not recommended to be used in prolonged high thermal exposure to due to it generates the degradation of the resin, which possibly results in a loss of mechanical and fracture strength. These drawbacks are limiting the extensive use of PC as becoming not more efficient option [18,19].

Several studies have been undertaken to improve the PC performance throughout the insertion of flame retardant, coupling agent, and fibers, or by the replacement of the solid part by industrial by-products (i.e., industrial wastes, recycle aggregates, etc.) [2,4,20–23]. However, few studies have been performed to investigate the retaining of mechanical and fracture properties of PC after long exposure in extreme thermal conditions. Elalaoui et al. showed that the mechanical and physical PC containing epoxy was obviously influenced by thermal conditions but still more efficient than Portland cement concrete even after exposure to temperatures around 250 °C [13]. Apart from the mechanical and fracture properties, PC composite incorporating epoxy as main binder material are relatively more expensive than composite made with other thermosets such as polyesters [9,24]. Regardless their high cost ranging between three and fifteen times more than other thermosets, epoxy resins are preferred in special applications where high cost can be easily balanced by their overall superior mechanical and fatigue properties [7,11,25,26].

The inclusion of synthetic or natural fibers seems to be an effective option to overcome the PC shortcoming and increase the strength capacity, ductility, and toughness of PC [27,28]. G.Martínez et al. [27] investigated the effects of adding polypropylene fibers on compressive properties of polymer concrete. Their results revealed that compressive of PC reinforced with fibers was improved. Reis et al. studied the effect of reinforcing epoxy PC by chopped glass (inorganic) or carbon fibers (organic) with different proportions [11,29]. A similar approach was taken by Rokbi et al. [30] who tried to valorize the use of natural resources by using laminated vegetable fibers consisting of woven fabric jute in various orientations to lead up the reduction in the environmental impact and improve the mechanical properties of PC. In both studies, it was shown clearly that the fractures, elasticity modulus, compressive, and splitting tensile and flexural strength of PC reinforced with fibers were improved.

Many models have been used to predict the crack propagation for different materials [29] such as non-linear elastic fracture mechanics, linear elastic fracture mechanics (LEFM), fictitious crack model (FCM), effective crack model (ECM), cohesive crack model (CCM), size-effect model, and two-parameter model (TPM). It was reported that the carbon fiber reinforcement improves the toughness of the PC by 29%, while the glass fibers generate a smaller improvement of 13%. The tenacity of PC is 36% higher than that of Portland cement concrete which is between 0.74 and 1.53 [11,30]. The last finding indicates that the PC is more resistant to crack widening even without reinforcements.

The TPM developed by Jenq and Shah [31] and based on determining the change in potential energy when crack extents will be used in this study as it is considered as one of the fracture models to implement [11].

A review of recent literature showed a considerable number of studies on polymer PC. However, limited work has been conducted to investigate the mechanical and fracture properties of fiber reinforced polymer concrete, which undergoes extreme thermal exposure conditions. Hence, the main objective of this study is to evaluate the degradation and the retention of the residual mechanical characteristics and toughness of epoxy-based polymer concrete reinforced randomly with short carbon and polypropylene fibers after exposure to elevated temperatures. In this study, short fibers were added in a rate of 1 and 2% by the total weight of the PC composite. The exposure temperatures cycles used were ranged between 23 to 250 °C. Various tests were conducted including flexure strength, elastic modulus, toughness, tensile crack opening, density, and porosity to evaluate the effect of addition of short fibers on fracture properties of PC.

2. Materials, Sample Preparation and Methods

2.1. Materials

A thermosetting two-component product (hardener and resin) was selected to prepare the polymer concrete. Commercialized epoxy resin under the reference EPONAL 371 obtained from Bostik SA (France) was used in this study. The epoxy was cross-linked with chemical modified polyamines hardener. It was mixed immediately before use for common applications. Advantageously, the curing of the selected thermosetting resin can be implemented at a temperature at 10–35 °C, but preferably at room temperatures (20–25 °C) as prescribed. The service temperature range of this epoxy has not been mentioned in technical data but the glass transition temperature was 82.51 °C, as measured using a differential scanning calorimetry machine TA Instruments Q100 DSC (Waters LLC, New Castle, UK). The epoxy properties are presented in Table 1.

Table 1. Properties of used materials according to the manufacturer.

Properties	EPONAL 371	Polypropylene Duomix® Fire M6	Aceca® ECO-H6
Tensile strength (MPa)	31.7 ± 3.2	300	3530
Young's modulus (MPa)	3800 ± 130	3500–3900	230
Elongation at break (%)	1.2 ± 03	15	1.7
Compressive strength (MPa)	81.05 ± 8.9	–	–
Length (mm)	–	6	3
Density at 23 °C (g/cm^3)	1.42–1.48	0.91	1.73–1.96
Color	Beige	transparent white	black
Brookfield viscosity at 23 °C (Pa.S)	5-12	–	–
Filament diameter (μm)	–	18	7
Melting temperature (°C)	–	160-165	3500

Two kinds of fibers (carbon and polypropylene) were chosen to be embedded into the PC's mixes (Table 1). A carbon fiber marketed as Aceca® ECO-H6 was supplied by Torayca Company, Japan. These fibers (having cylindrical shapes) are chopped into short lengths in the interest to help dispersion in viscous fluids. The second type of fiber, labeled under the name of DUOMIX FIRE® (M6), was provided by Bekaert (Belgium) as fine monofilaments polypropylene. These fibers have a tubular shape and come in the form of clusters which can be easily dispersed during mixing. The main disadvantage of these types of fibers was lowering the workability of the mix for the same volume of polymer.

Sand graded from 0 to 4 mm and a gravel graded from 4 to 10 mm were used as fine and coarse aggregates, respectively. The specific gravity of the sand was 2.47, whereas the specific gravity of gravel was 2.53 g/cm^3. Preliminary investigations based on the compressible packing model (CPM) was conducted to obtain the proportions of aggregates that guaranteed the maximum compactness of solid grains while minimizing the amount of polymer needed to wet fully the aggregates and fill gaps between. Portland cement CEM I 52.5 N having a specific density of 3.13 g/cm^3 and a specific surface equal to 3590 cm^2/g as evaluated via Blaine apparatus was also used in this study.

2.2. Sample Preparation

In this study, four PC mixes with fibers reinforcements and 13% of polymer, by weight of PC, were prepared. The PC samples were prepared according to a process described in the work of Elalaoui et al. [13].

The content of polymer was kept constant for the all mixes. This content was selected as the optimal amount in the basis of previous studies [13,25,32–35]. The fibers were added at the rate of 1% and 2%, by the total weight of the polymer. The PC compositions are listed in Table 2. The formed PC mixes are named as PC-Xp, where Xp% denotes the type of reinforcement (C for carbon and P for polypropylene) and p referring to the mass fraction of the introduced fibers in term of percentage. The PC was casted according to a well-defined process as described in a previous published study [13]. Ordinary Portland cement concrete (OCC) samples were also casted by mixing the same aggregates proportions for comparison purposes. The OCC mix constituent is shown in Table 2.

Table 2. PC and OCC mixes with the varying fiber's content.

Material	Concrete Mix				
	PC	PC-C1	PC-C2	PC-P1	OCC
Polymer (%)	13	13	13	13	-
Cement CEM I 52.5 (%)	–	–	–	-	15.6
Sand (%)	56.6	55.6	54.6	55.6	35.1
Gravel (%)	30.4	30.4	30.4	30.4	41
Water	–	–	–	–	8.3
Carbon fibers (%)	–	1	2	–	–
Polypropylene fibers (%)	–	–	–	1	–

2.3. Testing Method

Several tests were conducted including compression and flexure strength, elastic modulus determined using the ultrasonic pulse velocity method (UPV), toughness, tensile crack opening, density, and porosity (using mercury intrusion porosimetry, MIP) to evaluate the effect of addition of short fibers on fracture properties of PC. The principle of MIP is to inject a no-wetting liquid such as mercury inside the pores of a small sample (10 cm^2 as maximum) under vacuum conditions and under pressure ranging from 14 to 414 MPa.

A uniaxial compression tests were carried out using cylinders sample (50 mm × 100 mm). The effect of adding fibers on the flexure strength, toughness, and mechanical residual properties of PC were investigated by performing three-point bending test beam (50 mm × 50 mm × 305 mm). These tests were conducted in a displacement-control mode with a constant rate of 1.0 mm/min and 1.25 mm/min for bending and compression, respectively, as per the procedure stated in RILEM PCM8 1995 [36].

The midpoint deflection of beams was evaluated by the mean of a linear variable differential transformer (LVTD) rested on a steel L-shape bracket attached to specimens at mid-height as shown in the figure below (Figure 1).

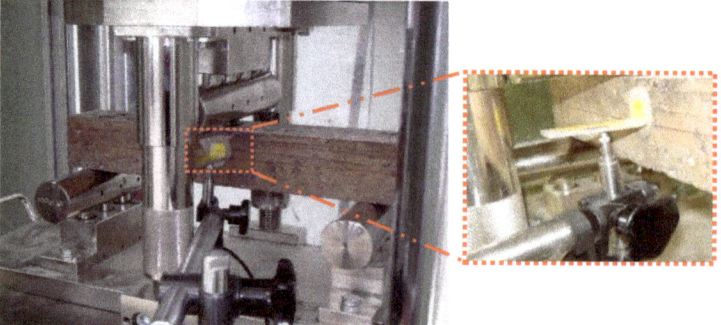

Figure 1. Details of bending test apparatus.

Concrete toughness was evaluated throughout pre-notched beams using a 250 kN closed loop INSTRON machine. The crack mouth opening displacement (CMOD) was measured by virtue of clip-gauge attached to a knife-edges installed at the bottom of the beams (Figure 2). The mid-span deflection was recorded using an LVDT placed at the left side at the bottom of the specimens. The dimensions of sample and central U-shape notch properties are itemized in Table 3.

Figure 2. Experimental setup for CMOD controlled three-point bending test.

Table 3. Dimensions of sample and central notch properties.

	Beam Dimensions $b \times d \times L$	Rete of Loading (mm/min)	Span Length S	Length of the Notch a_0	Width of the Notch	$\frac{a_0}{d}$
PC	50 × 70 × 480	0.05	400	14	3 max	0.200
OCC	80 × 150 × 750	0.05	640	48	3 max	0.312

For the OCC mix, beam samples of 70 mm × 70 mm × 280 mm for bending test and cylinder samples of 150 mm × 300 mm for compression sample were casted. The OCC beams samples were loaded at rate of 0.05 MPa/s, whereas the cylinder samples was loaded at 0.5 MPa/s.

The fracture toughness values of PC in mode-I, K_{IC}, and the fracture energy G_F were calculated using a two fracture parameters model [37,38] which appears to give rather realistic prediction of concrete fracture behavior. An actual crack is replaced by an equivalent fictitious crack [39].

The effective crack length and the peak load, are required to determine K_{Ic} using linear elastic fracture mechanics, LEFM. This value of K_{Ic} was found to be independent of the specimen size [40].

The stress intensity factor is given by means of Equation (1):

$$K_{Ic} = \sigma_{NC} \sqrt{\pi a_c} \, F_1(\alpha) \tag{1}$$

where $\sigma_{NC} = \frac{3F.S}{2b.d^2}$ and $\alpha = \frac{a_c+d_0}{d+d_0}$

$$F_1(\alpha) = \frac{1.83 - 1.85\alpha + 4.76\alpha^2 - 5.3\alpha^3 + 2.51\alpha^4}{(1+2\alpha)(1-\alpha)^{\frac{3}{2}}} \tag{2}$$

The critical crack length is presented by Equation (3)

$$a_c = a_0 \frac{C_u \, f_2(\alpha_0)}{C_i \, f_2(\alpha)} \quad \text{and} \quad \alpha_0 \frac{a_0}{d} \tag{3}$$

The geometrical parameters S, b, d, a_0 and d_0 are described in Figure 3.

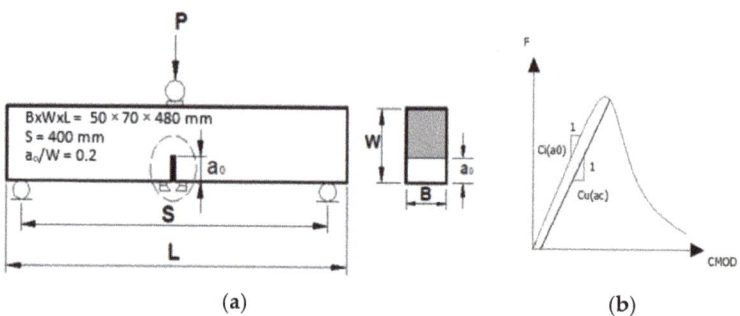

Figure 3. Determination of fracture parameters of concrete (**a**) geometrically characteristic values, (**b**) typical F-CMOD curve.

The fracture energy is calculated according to the following relation as postulated by the following LEFM equation:

$$K_{Ic} = \sqrt{E.G_F} \qquad (4)$$

The prepared specimens were heated at a rate of 0.5 °C/min until target temperature was reached according to the predefined protocol. The temperature was maintained for 3 h before being decreases with the same heating rate till samples reached the ambient temperature being kept for 24 h until the specimens were tested. To ensure that the setting temperature (from 100 °C to 250 °C) is compatible with specimen temperature, two k-type thermocouples mounted on data logger were attached to one side and in the middle of samples for a continuous recording (Figure 4).

Figure 4. Temperatures control through installation of thermocouples.

A set of four specimens was allocated for each test type. The tests were conducted on concretes allowed to cure for 7 days at prescribed room temperature for both unreinforced and reinforced PC and 28 days age for OCC.

The setting and recorded temperatures were close and only a difference not exceeding ±3 °C was reported (Figure 5).

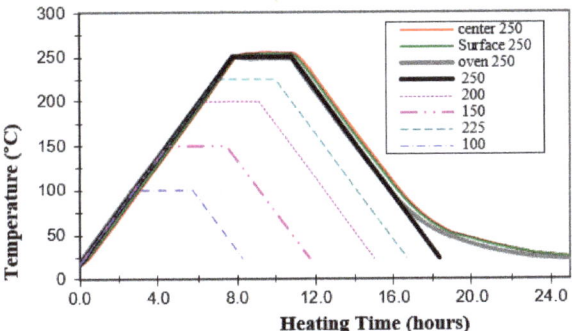

Figure 5. Real time-Experimental temperature records vs. heating time.

3. Results and Discussion

3.1. Fibers Adding Effects on the Density and Porosity Properties of PC

The effect of addition of short fibers on the density PC mixed is shown in Table 4. The results reveals that the inclusion of short fibers have a slight effect on the density of PC concrete. The density of the studied mixes ranged from 2.16 g/cm^3 to 2.23 g/cm^3. In fact, a marginal decrease was observed for the entire mixes. The same trend can be observed for the carbon group of mixes as the density decreased with higher fiber dosages. The maximum difference in densities recorded over the mixes was 3.2%, which shows that incorporation of fibers with small quantities did not significantly affect the concrete density. In addition, it was noted that the density of polypropylene fibers is lower than that of polymer. This could be due to the deficiency of uniform distribution during the mixing process and thereby in concrete bulk.

Table 4. Density and porous structure properties of studied concretes.

Concrete System	Density (g/cm^3)	Total Porosity (%)	d_c (µm)
PC	2.23	3.6	175.4
PC-C1	2.20	4.6	83.8
PC-C2	2.17	8.8	51.9
PC-P1	2.16	12.30	61.0
OCC	2.37	15.13	–

The pore structures (i.e., mean pore size, pore size distribution, and various pores size proportions) were characterized by means of MPI and results are presented in Table 4. The introduction of fibers leads also to a decrease in the mean most distributed pore diameter d_c. The increase in the porosity of PC reinforced with polypropylene fibers can be attributed to volume expansion of the fibers during mixing which conducts to an additional porosity. In general, the addition of short fibers to polymer concrete generates difficulties of the solid particles compaction causing an increase in the total porosity and a decrease in the most distributed pore diameter, d_c as shown in Figure 6. Similar phenomena have been observed by Bentur et al. [41] and Barbuta et al. [42]. Although, the porosity of PC mixes increases with fiber content, but is still significantly lower than that of OCC samples.

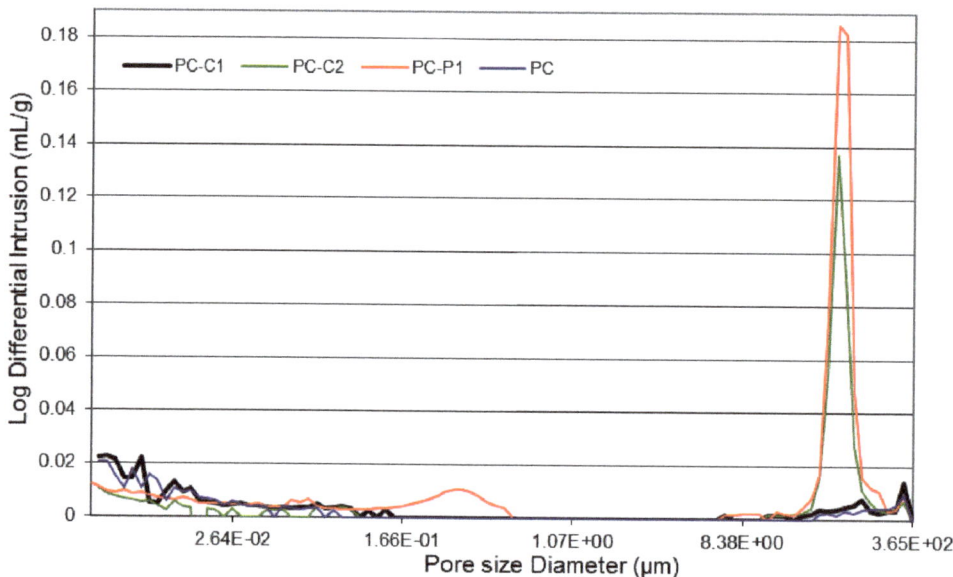

Figure 6. Pores size distribution for different PC mixes.

3.2. Effects of Adding Fibers on the Modulus of Elasticity of PC Systems

The dynamic Young's modulus E_d and the shear modulus μ_d are determined using UPV. The two dynamic modulus are given by the following Equations (5) and (6):

$$V_L = \left(\frac{E(1-\vartheta)}{\rho(1+\vartheta)(1-2\vartheta)} \right)^{\frac{1}{2}} \quad (5)$$

$$\mu_d = \rho\, V_T^2 \quad (6)$$

The value of Poisson's ratio is given by Equation (7):

$$\vartheta = \frac{1 - 2(V_T/V_L)^2}{2 - 2(V_T/V_L)^2} \quad (7)$$

In the Equation (5) to Equtaion (6), E_d represents the dynamic elasticity modulus (MPa), V_L is the compressive P-wave velocity, V_T is shear wave velocity (km/s), ν is Poisson's ratio, and ρ is the density (kg/m^3).

The results of the experimental program revealed that the static elastic modulus (Es) decreased slightly as the fiber content increased (Figure 7a,b). This marginal variation is the result of two balanced phenomena: an increase due to the high elastic modulus of the carbon fibers and decrease due to the increase in porosity and the anisotropic feature. By increasing the fraction of fibers, the growth of porosity outweighs the contribution of fibers. The lowest rigidity is observed for PC-P1 because polypropylene fibers have a low elastic modulus similar to that of the epoxy binder. Moreover, PC-P1 has higher values of the total porosity compared to control sample. In the same context, Reis and Kumar [10,43] in their research paper confirmed that adding carbon and glass fibers do not improve the compressive elastic modulus of the composites but in opposite slight decrease was observed while carbon fiber reinforcements are incorporated.

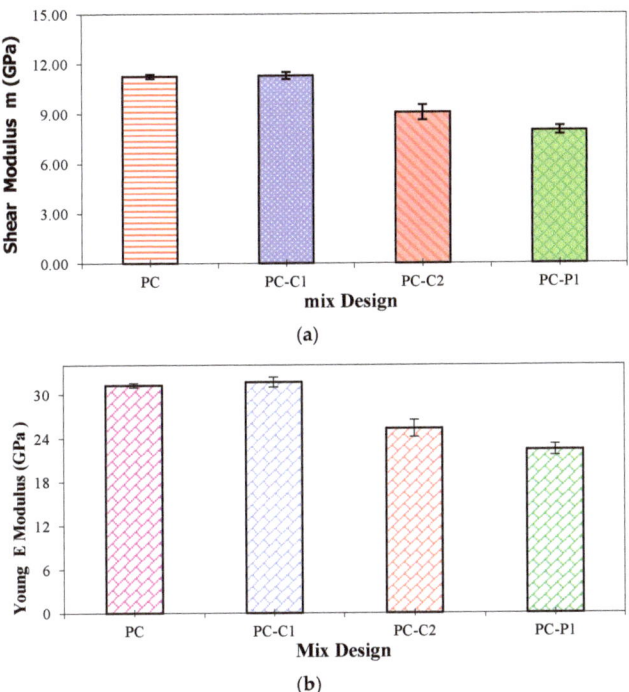

Figure 7. Elastic characteristics of polymer concretes: (**a**) Shear Modulus, (**b**) Young Modulus.

By adding fibers to PC, all mixes showed indistinctly an increase in the modulus with sample age and being stabilized after a period of 5 to 7 days (Figure 8). This can be attributed to the progress of the curing reaction taken end at the 7 days as maximum leading to the formation of polymeric epoxy structures [16]. It should be mentioned that the modulus of elasticity records for unreinforced PC displayed are in accordance with the values described in literature for epoxy-based PC [11,44].

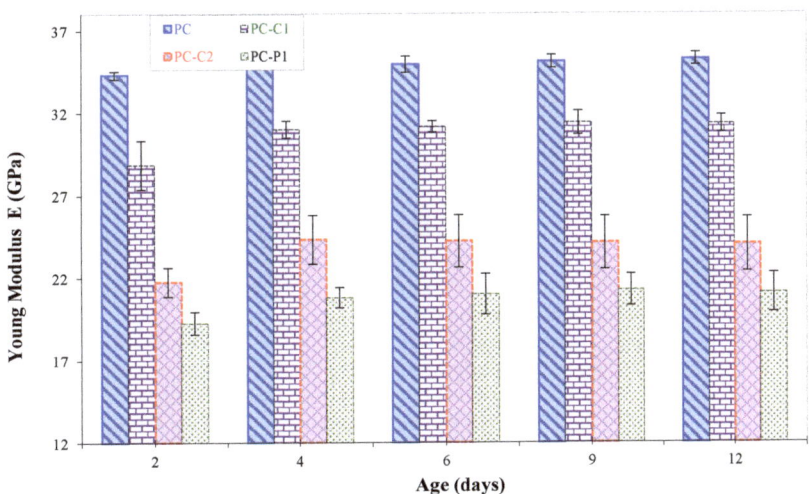

Figure 8. Elastic modulus vs. age for PC mixes.

3.3. Fibers Adding Effects on the Mechanical Properties of PC

The results of flexural and compressive strength of PC mixes are presented in Figure 9a,b. The composite systems exhibit lower flexural and compressive characteristics when reinforced with polypropylene and carbon fibers, regardless of the type of reinforcements. In fact, it was reported that polypropylene fibers as example show weak binder–fiber contacts and this decrease the performance of the final product, regardless of the binder type [45].

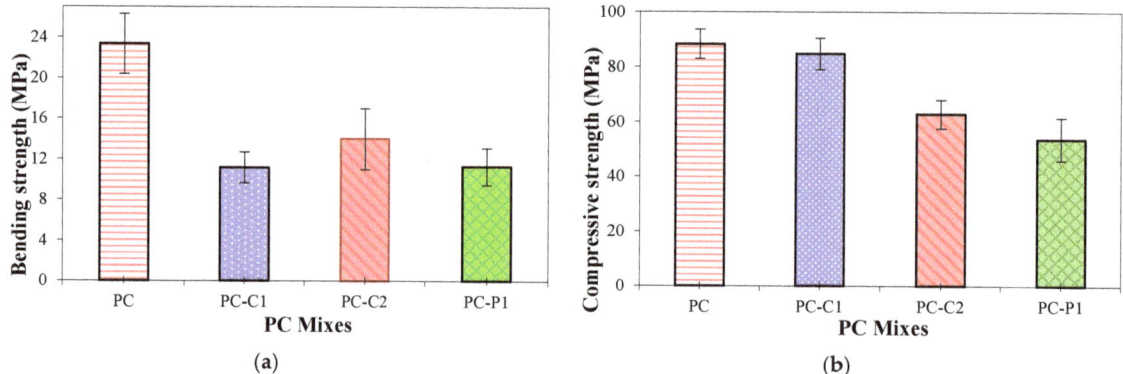

Figure 9. Bending and compressive resistances of different PC sets: (**a**) Flexural strength, (**b**) Compressive strength.

However, the results show that flexural strength of PC made with carbon fibers increased with the increase in fibers content (Figure 9a). In average, flexural strengths increased 20% approximately for 2% fiber content, and compressive strength decreased 25.9% for the same fiber content, comparing to PC-C1.

The decrease in the mechanical resistances is probably due to poor adhesion between the fibers and the matrix and to the high values of porosity. This decrease is a sign of decrease in transverse bonds in polymer, which decrease the stiffness and increase ductility of polymer concrete [46]. In fact, adding 1% of carbon fibers does not affect the compressive strength and the elastic characteristics but decreases the flexure strength. In other words, fibers do not increase PC strength but their addition to the mixture diminished the signs of brittleness behavior of unreinforced polymer concrete [20], which is translated by a change in bending behavior from quasi-linear brittle to nonlinear ductile, a nonlinearity heightened by increasing the fiber content.

It can be concluded that the addition of fibers did not accomplish the expected reinforce or at least has the same strength characteristics as unreinforced PC [17]. Hence, only the toughness of polymer concretes reinforced with carbon fibers will be considered thereafter.

3.4. Fibers Adding Effects on Toughness at High Temperatures

Figure 10a,b represent, respectively, the variation of stress intensity factor K_{IC} and the fracture energy G_F as a function of the exposure temperature for both the PC system and the OCC. It can be seen that PC is more resistant to crack propagation than OCC. By increasing the exposure temperature, the toughness is improved up to 225 °C for both concrete before declining slightly at higher temperatures (250 °C). Such a variation has been similarly observed for the flexural strengths mainly in the case of plain PC as a result of action of two coexistent and competitive phenomena: (i) post crosslinking initiated by heat and (ii) thermo-oxidative degradation of the epoxy polymer [47]. This variation is emphasized by a loss of bond between aggregates and the binder as demonstrated in a previous work [14].

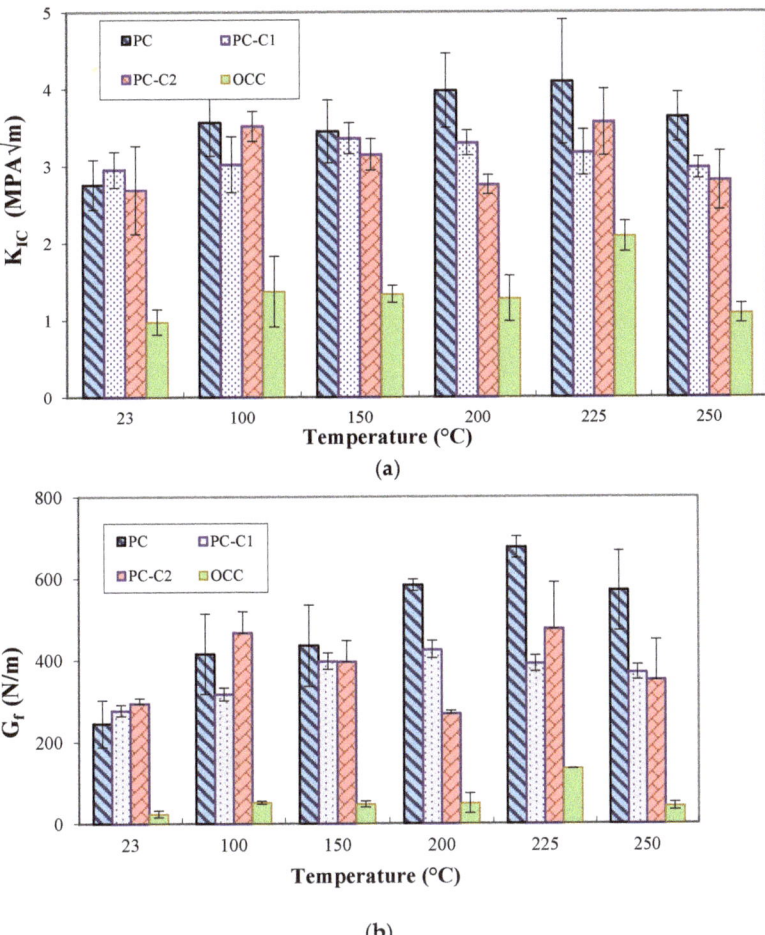

Figure 10. Fracture parameters as a function of heating temperature for concrete mixes. (**a**) Stress intensity factor (**b**) Fracture energy.

The effect of carbon fiber introduction on the fracture properties of PC is studied and the results are depicted on Figure 10a,b. It can be seen that adding 1% of carbon fibers to polymer concretes results in an enhancement on its fracture properties at room temperature, but this effect is vanished by increasing the exposure temperature as a result of the decrease in bond strength in the fiber–matrix interface while rising temperature [14] added to the other types of degradation mentioned earlier.

Figure 11 shows the load–CMOD graph for the tested notched beams as a function of heating temperature. The overall load–CMOD curves have presented a smooth softening curves after the peak loads which indicates that tests were conducted under a stable test regime. All curves observed show a similar general trend for all specimens.

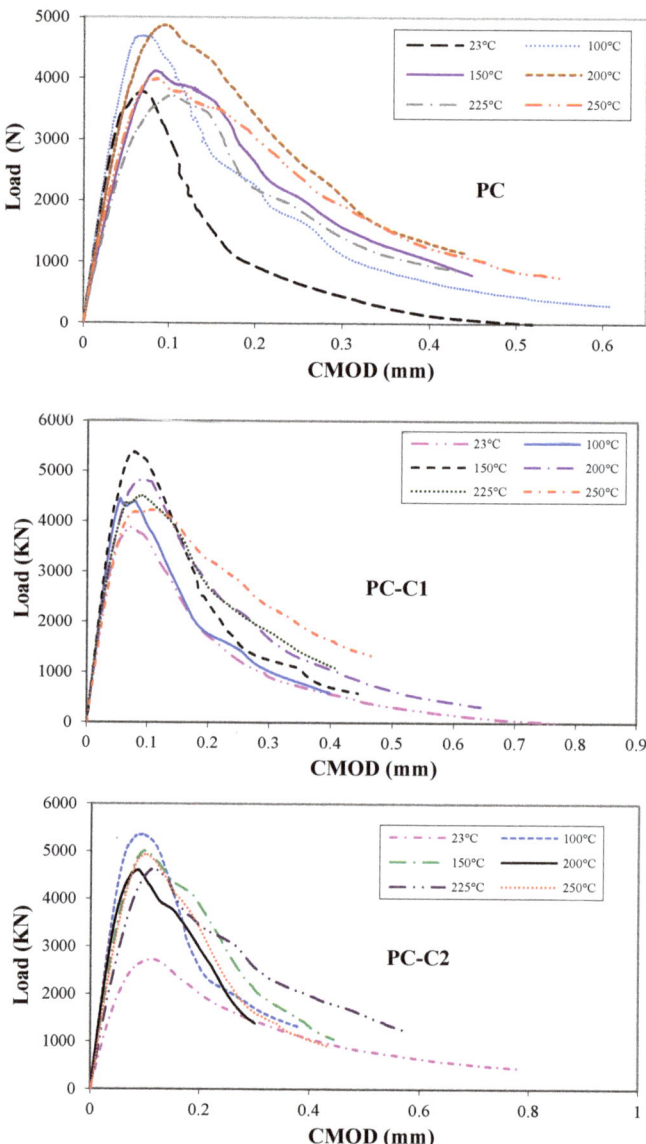

Figure 11. Load–CMOD for PC systems for different heating temperatures.

It can be seen also that for reinforced PC mixes the load carrying capacity increases until 150 °C and then decreases for higher temperatures. Moreover, a significant change in the post-peak behavior was observed, the ductile behavior becomes less brittle [48]; similar to the type of behavior observed in non-reinforced concrete beams. This ductility enhancement yields slower crack propagation, a phenomenon attributed to fracture energy increase due to the existence of a micro-cracks network. It should be highlighted that more energy dissipation is necessary for the coalescence of micro-cracks into a single macro-crack [49]. The use of fine and short fibers in composite helps to reduce their critical crack length and increase fiber–matrix interface area to intercept cracks. This means that if crack passes through fibers (being less stiff than fiber), it will encounter the resin which will control and slow crack. If this does not stop crack propagation, the next fiber can intercept crack and stop it. Fibered PC force cracks to follow a very devious path that needs larger amounts of energy to create new fracture surfaces.

4. Conclusions

The effect of incorporating short carbon and polypropylene fibers combined with the exposure to high heating temperatures reaching 250 °C on the mechanical and fracture properties of epoxy polymer-based concrete is investigated in depth in this study and the following remarks were derived:

- Epoxy-based concrete possesses higher mechanical properties compared to ordinary cement concrete.
- For temperatures less than 250 °C, the epoxy polymer concrete is still more efficient than ordinary cement concrete.
- The addition of short carbon fibers content by a rate of 1% by weight to polymer concrete indicated that there was no indicative difference in the concrete density, the elastic characteristics, the compressive strength, and led to a decrease in the flexural strength. It also did not enhance the fracture properties at room temperature.
- When exposed to high temperatures of up to 150 °C, the 1% fiber introduction resulted in a load-carrying capacity increase. This enhancement of fracture properties is vanished for higher heating temperatures.
- Compared to the post cracking behavior at room temperature, the ductility of 1%-fibered polymer concrete increases when it is exposed to high temperatures, resulting in slower crack propagation.
- Carbon fibers introduction at content of 2% by weight of polymer-based concrete did not improve its mechanical and fracture properties. The same was observed for polypropylene fibers used with a fraction of 1% by weight.

5. Recommendations

Due to complexity and co-existence of many phenomena, deeper investigations are hence needed to explain their interaction that affect the properties of reinforced fiber polymer concrete. Curing conditions of samples and their effects in the fracture and thermo-mechanical properties seems to be interesting to be investigated and evaluated to build a complete and consistent knowledge about polymer concrete exposed to extreme in-service conditions.

Funding: The author would like to thank deanship of Scientific Research at Majmaah University for supporting this work under Project number No. R-2023-17.

Institutional Review Board Statement: Not applicable.

Informed Consent Statement: Not applicable.

Data Availability Statement: Not applicable.

Acknowledgments: The author would like to thank deanship of Scientific Research at Majmaah University for supporting this work under Project number No. R-2023-17.

Conflicts of Interest: The authors declare no conflict of interest.

References

1. Ohama, Y. Recent progress in concrete polymer composites. *Cem. Concr. Compos.* **1997**, *5*, 31–40. [CrossRef]
2. Alperen, B.H.; Sahin, R.A. Study on mechanical properties of polymer concrete containing electronic plastic waste. *Compos. Struct.* **2017**, *178*, 50–62.
3. Douba, A.; Emiroglu, M.; Kandil, U.F.; Taha, M.M.R. Very ductile polymer concrete using carbon nanotubes. *Constr. Build. Mater.* **2019**, *196*, 468–477. [CrossRef]
4. Awham, M.H.; Mohammad, T.H. Characteristics of polymer concrete produced from wasted construction materials. *Energy Procedia* **2019**, *157*, 43–50.

5. Toufigh, V.; Toufigh, V.; Saadatmanesh, H.; Ahmari, S.; Kabiri, E. Behavior of polymer concrete beam/pile confned with CFRP sleeves. *Mech. Adv. Mater. Struct.* **2019**, *26*, 333–340. [CrossRef]
6. Li, H.; Sun, J.; Chen, J.; Cai, S.; Yin, J.; Xu, J. Development Status and Application of Polymer Concrete. In Proceedings of the 2016 International Forum on Energy, Environment and Sustainable Development (Advances in Engineering Research), Shenzhen, China, 6–17 April 2016. [CrossRef]
7. Rahman, M.; Akhtarul, I.M. Application of epoxy resins in building materials: Progress and prospects. *Polym. Bull.* **2021**, *79*, 1949–1975. [CrossRef]
8. American Concrete Institute Committee. *Guide for Polymer Concrete Overlays*; ACI Committee Report ACI 548.5R-16; American Concrete Institute Committee: Farmington Hills, MI, USA, 2016.
9. Kumar, R. A review on epoxy and polyester based polymer concrete and exploration of polyfurfuryl alcohol as polymer concrete. *J. Polym.* **2016**, *2016*, 7249343. [CrossRef]
10. Kumar, M.; Mohan, D. Studies on polymer concretes based on optimized aggregate mix proportion. *Eur. Polym. J.* **2004**, *40*, 2167–2177. [CrossRef]
11. Gao, Y.; Zhang, H.; Huang, M.; Lai, F. Unsaturated polyester resin concrete: A review. *Constr. Build. Mater.* **2019**, *228*, 116709. [CrossRef]
12. Aliha, M.R.M.; Karimi, H.R.; Abedi, M. The role of mix design and short glass fiber content on mode-I cracking characteristics of polymer concrete. *Constr. Build. Mater.* **2022**, *317*, 126139. [CrossRef]
13. Elalaoui, O.; Ghorbel, E.; Mignot, V.; Ouezdou, M.B. Mechanical and physical properties of epoxy polymer concrete after exposure to temperatures up to 250 °C. *Constr. Build. Mater.* **2012**, *27*, 415–424. [CrossRef]
14. Ataabadi, H.S.; Sedaghatdoost, A.; Rahmani, H.; Zare, A. Microstructural characterization and mechanical properties of lightweight polymer concrete exposed to elevated temperatures. *Constr. Build. Mater.* **2021**, *311*, 125293. [CrossRef]
15. Gao, X.; Han, T.; Tang, B.; Yi, J.; Cao, M. Reinforced Structure Effect on Thermo-Oxidative Stability of Polymer-Matrix Composites: 2-D Plain Woven Composites and 2.5-D Angle-Interlock Woven Composites. *Polymers* **2022**, *14*, 3454. [CrossRef] [PubMed]
16. Kodur, V.K.R.; Bhatt, P.P.; Naser, M.Z. High temperature properties of fiber reinforced polymers and fire insulation for fire resistance modeling of strengthened concrete structures. *Compos. Part B Eng.* **2019**, *175*, 107104. [CrossRef]
17. Khotbehsara, M.M.; Manalo, A.; Aravinthan, T.; Reddy, K.R.; Ferdous, W.; Wong, H.; Nazari, A. Effect of elevated in-service temperature on the mechanical properties and microstructure of particulate-filled epoxy polymers. *Polym. Degrad. Stab.* **2019**, *170*, 108994. [CrossRef]
18. Elalaoui, O.; Ghorbel, E.; Ouezdou, M.B. Influence of flame retardant addition on the durability of epoxy based polymer concrete after exposition to elevated temperature. *Constr. Build. Mater.* **2018**, *192*, 233–239. [CrossRef]
19. Ribeiro, M.C.S.; Novoa, P.R.; Ferreira, A.J.M.; Marques, A.T. Flexural performance of polyester and epoxy polymer mortars under severe thermal conditions. *Cem. Concr. Compos.* **2004**, *26*, 803–809. [CrossRef]
20. Wang, Y. Fiber and Textile Waste Utilization. *Waste Biomass Valoriz.* **2010**, *1*, 135–143. [CrossRef]
21. Kou, S.C.; Poon, C.S. A novel polymer concrete made with recycled glass aggregates, fly ash and metakaolin. *Constr. Build. Mater.* **2013**, *41*, 146–151. [CrossRef]
22. Farooq, M.; Banthia, N. Strain-hardening fiber reinforced polymer concrete with a low carbon footprint. *Constr. Build. Mater.* **2022**, *314*, 125705. [CrossRef]
23. Sosoi, G.; Barbuta, M.; Serbanoiu, A.A.; Babor, D.; Burlacu, A. Wastes as aggregate substitution in polymer concrete. *Procedia Manuf.* **2018**, *22*, 347–351. [CrossRef]
24. Reis, J.M.L.; Jurumenh, M.A.G. Experimental investigation on the effects of recycled aggregate on fracture behavior of polymer concrete. *Mater. Res.* **2011**, *14*, 326–330. [CrossRef]
25. Ferdous, W.; Manalo, A.; Wong, H.S.; Abousnina, R.; AlAjarmeh, O.S.; Zhuge, Y.; Schubel, P. Optimal design for epoxy polymer concrete based on mechanical properties and durability aspects. *Constr. Build. Mater.* **2020**, *232*, 117229. [CrossRef]
26. Hashemi, M.J.; Jamshidi, M.; Aghdam, H.J. Investigating fracture mechanics and flexural properties of unsaturated polyester polymer concrete (UP-PC). *Constr. Build. Mater.* **2018**, *163*, 767–775. [CrossRef]
27. Martínez-Barrera, G.; Vigueras Santiago, E.; Hernandez Lopez, S.; Gencel, O.; Ureña-Nuñez, F. Polypropylene Fibers as Reinforcements of Polyester-Based Composites. *Int. J. Polym. Sci.* **2013**, *2013*, 143894. [CrossRef]
28. Shankar, R.S.; Srinivasan, S.A.; Shankar, S.; Rajasekar, R.; Kumar, R.N.; Kumar, P.S. Review article on wheat flour/wheat bran/wheat husk based bio composites. *Int. J. Sci. Res. Publ.* **2014**, *4*.
29. concrete reinforced with short carbon and glass fibers. *Constr. Build. Mater.* **2004**, *18*, 523–528. [CrossRef]
30. Rokbi, M.; Rahmouni, Z.; Baali, B. Performance of polymer concrete incorporating waste marble and alfa fibers. *Adv. Concr. Constr.* **2017**, *5*, 331–343.
31. Niaki, M.H.; Ahangari, M.G.; Izadi, M.; Pashaian, M. Evaluation of fracture toughness properties of polymer concrete composite using deep learning approach. *Fatigue Fract. Eng. Mater. Struct.* **2022**, *46*, 603–615. [CrossRef]
32. Dastgerdi, A.S.; Peterman, R.J.; Savic, A.; Riding, K.; Beck, B.T. Prediction of splitting crack growth in prestressed concrete members using fracture toughness and concrete mix design. *Constr. Build. Mater.* **2020**, *246*, 118523. [CrossRef]
33. Shokrieh, M.M.; Heidari-Rarani, M.; Shakouri, M.; Kashizadeh, E. Effects of thermal cycles on mechanical properties of an optimized polymer concrete. *Constr. Build. Mater.* **2011**, *25*, 3540–3549. [CrossRef]

34. Jafari, K.; Tabatabaeian, M.; Joshaghani, A.; Ozbakkaloglu, T. Optimizing the mixture design of polymer concrete: An experimental investigation. *Constr. Build. Mater.* **2018**, *167*, 185–196. [CrossRef]
35. Bedi, R.; Chandra, R.; Singh, S.P. Mechanical properties of polymer concrete. *J. Compos.* **2013**, *12*, 948745. [CrossRef]
36. RILEM PCM8, T.C.T. *Method of Test for Flexural Strength and Deflection of Polymer-Modified Mortar. Symposium on Properties and Test Methods for Concrete–Polymer Composites*; Technical Committee TC-113: Oostende, Belgium, 1995.
37. Mekonen, A. *Fracture Mechanics in Concrete and its Application on Crack Propagation and Section Capacity Calculation*; Addis Ababa University: Addis Ababa, Ethiopia, 2018.
38. Vantadori, S.; Carpinteri, A.; Fortese, G.; Ronchei, C.; Scorza, D. Mode I fracture toughness of fiber-reinforced concrete by means of a modified version of the Two-Parameter Model. *Proc. Struct. Integr.* **2016**, *2*, 2889–2895.
39. Yahya, M.A. *The Effect of Polymer Materials on the Fracture Characteristics of High Performance Concrete (HPC)*; Edinburgh Napier University: Edinburgh, UK, 2015.
40. Golewski, G.L. Fracture Performance of Cementitious Composites Based on Quaternary Blended Cements. *Materials* **2022**, *15*, 6023. [CrossRef]
41. Bentur, A.; Mindess, S. *Fibre Reinforced Cementitious Composites. Modern Concrete Technology Series*; Taylor and Francis: New York, NY, USA, 2007.
42. Barbuta, M.; Serbanoiu, A.A.; Teodorescu, R.; Rosca, B.; Mitroi, R.; Bejan, G. Characterization of polymer concrete with natural fibers. *Mater. Sci. Eng.* **2017**, *246*, 012033. [CrossRef]
43. Reis, J.M.L. Mechanical characterization of fiber reinforced Polymer Concrete. *Mater. Res.* **2005**, *8*, 357–360. [CrossRef]
44. Niaki, M.H.; Ahangari, M.G. *Polymer Concretes: Advanced Construction Materials*, 1st ed.; CRC Press: Boca Raton, FL, USA, 2022. [CrossRef]
45. Graziano, A.; Dias, O.A.T.; Petel, O. High-strain-rate mechanical performance of particle- and fiber-reinforced polymer composites measured with split Hopkinson bar: A review. *Polym. Compos.* **2021**, *42*, 4932–4948. [CrossRef]
46. Ahmad, W.; Khan, M.; Smarzewski, P. Effect of Short Fiber Reinforcements on Fracture Performance of Cement-Based Materials: A Systematic Review Approach. *Materials* **2021**, *14*, 1745. [CrossRef]
47. Elanchezhian, C.; Ramnath, B.V.; Hemalatha, J. Mechanical Behaviour of Glass and Carbon Fibre Reinforced Composites at Varying Strain Rates and Temperatures. *Procedia Mater. Sci.* **2014**, *6*, 1405–1418. [CrossRef]
48. Yunfu, O.; Zhu, D. Tensile behavior of glass fiber reinforced composite at different strain rates and temperatures. *Constr. Build. Mater.* **2015**, *96*, 648–656. [CrossRef]
49. Zhang, D.; Tamon, U.; Furuuchi, H. Fracture Mechanisms of Polymer Cement Mortar: Concrete Interfaces. *J. Eng. Mech. ASCE* **2013**, *139*, 167–176. [CrossRef]

Disclaimer/Publisher's Note: The statements, opinions and data contained in all publications are solely those of the individual author(s) and contributor(s) and not of MDPI and/or the editor(s). MDPI and/or the editor(s) disclaim responsibility for any injury to people or property resulting from any ideas, methods, instructions or products referred to in the content.

Article

Thermal Stabilities and Flame Retardancy of Polyamide 66 Prepared by In Situ Loading of Amino-Functionalized Polyphosphazene Microspheres

Wenyan Lv [1,2,*], Jun Lv [2], Cunbing Zhu [2], Ye Zhang [2], Yongli Cheng [2], Linghong Zeng [2], Lu Wang [2] and Changrong Liao [1,*]

[1] College of Optoelectronic Engineering, Chongqing University, Shazheng Road 174, Shapingba District, Chongqing 400044, China
[2] College of Materials Science and Engineering, Chongqing University of Technology, Hongguang Road 69, Banan District, Chongqing 400054, China
* Correspondence: lwy198585@cqut.edu.cn (W.L.); crliao@cqu.edu.cn (C.L.)

Abstract: The flame-retardant polyamide 66 composites (FR-PA66) were prepared by in situ loading of amino-functionalized polyphosphazene microspheres (HCNP), which were synthesized in the laboratory and confirmed by a Fourier transform infrared spectrometer (FTIR), scanning electron microscope (SEM), and transmission electron microscope (TEM). The thermal stabilities and flame retardancy of FR-PA66 were measured using thermogravimetric analysis (TGA), a thermogravimetric infrared instrument (TG-IR), the limiting oxygen index (LOI), the horizontal and vertical combustion method (UL-94), and a cone calorimeter. The results illustrate that the volatile matter of FR-PA66 mainly contains carbon dioxide, methane$_4$, and water vapor under heating, accompanied by the char residue raising to 14.1 wt% at 600 °C and the value of the LOI and UL-94 rating reaching 30% and V-0, respectively. Moreover, the addition of HCNP decreases the peak of the heat release rate (pHRR), total heat release (THR), mass loss (ML), and total smoke release (TSR) of FR-PA66 to 373.7 kW/m^2, 106.7 MJ/m^2, 92.5 wt%, and 944.8 m^2/m^2, respectively, verifying a significant improvement in the flame retardancy of PA66.

Keywords: PA66; polyphosphazene; microsphere; thermal stability; flame retardant

1. Introduction

Polyamide 66 is an important engineering plastic, mainly used in electronic and electrical, transportation, aerospace, and other fields. The LOI of PA66 is 24.0%, with a UL-94 V-2 rating. It is particularly important to improve the flame retardancy of PA66 [1]. Currently, flame retardants are usually added to improve the flame retardancy of PA66, such as halogenated flame retardants decabromodiphenyl ether [2], metal hydroxide [3], nano clay [4], nitrogen compounds [5], and phosphorus-containing organic compounds [6]. Because halogenated flame retardants are prone to produce toxic and corrosive smoke and gas during combustion, they easily corrode equipment, harm the environment, and harm the human body.

To avoid the fire hazards and environmental pollution accompanied by using halogen-based flame retardants in polyamide [7], phosphorus-containing flame retardants have been investigated and found to be effective substitutes with high efficiency, high charring, no melting drip, and economic performance [8]. These characteristics can improve the fire resistance and char formation of polyamide in the combustion process owing to their specific flame retardant mechanism [9]. However, the low degradation temperature of the elementary substance in phosphorus-type flame retardants limits their application in the processing of polyamide [10].

In recent years, polyphosphazene derivatives have been proven to be the most prospective flame retardants owing to their designability and char formation. For instance, novel cross-linked polyphosphazene microspheres were synthesized with 4,4'-dihydroxy biphenyl and tannic acid as co-monomers and decorated with layered double hydroxide to improve the flame retardancy of the epoxy resin. The results revealed that the EP containing 4.0 wt% microspheres exhibited the highest LOI of 29.7 and a UL-94 V-0 rating. Furthermore, the pHRR, THR, and TSR of the epoxy resin composites were significantly reduced and were superior to most of their previously reported counterparts [11]. Moreover, poly-(cyclotriphosphazene-co-4,4-sulfonyldianiline) (PDS) containing amino and hydroxyl groups was synthesized by Z.P Mao and co-workers [12], and the effect of polyphosphazene with different functional groups on the flame retardancy of polyethylene terephthalate was studied. After adding 5 wt% PDS, the LOI of PET composites increased to 33.1%, and they passed the UL-94 V-0 test.

In addition, a series of polyphosphazene flame retardants have been synthesized to improve the flame retardancy, thermal stability, and mechanical properties of polymers by Y Hu and co-workers [13]. For example, a novel allyl-functionalized linear polyphosphazene (PMAP) was designed and synthesized. With the inclusion of 3 wt% PMAP, the pHRR and TSP of composites were reduced by 51.3% and 17.8%, respectively, and the residual char increased significantly as well. Moreover, the impact strength increased by 85.3%, indicating that the toughness was effectively enhanced [14]. Another novel multifunctional organic–inorganic hybrid, melamine-containing polyphosphazene wrapped ammonium polyphosphate (PZMA@APP) with rich amino groups was prepared and used as an efficient flame retardant. The obtained sample passed the UL-94 V-0 rating with a 10.0 wt% addition of PZMA@APP. Notably, the inclusion of incorporating PZMA@APP led to a significant decrease in the fire hazards of EP (75.6% maximum decrease in pHRR and 65.9% maximum reduction in THR) [15].

Because of the excellent natural flame retardant synergies and thermal stability of polyphosphazene, it is believed that an increasingly important role will be played by polyphosphazene modified with nanoparticles in flame retardant applications. For instance, polyphosphazene loading with MoS_2 nanosheets has been successfully fabricated and significantly improved the flame retardancy of epoxy resin, i.e., 30.7% and 23.6% reductions in pHRR and THR, respectively [16]. Moreover, amino-functionalized carbon nanotubes/polyphosphazene hybrids (AFP@CNTs) were designed and synthesized to enhance the fire safety and mechanical properties of EP. With the addition of 1.5 wt% AFP@CNTs, the pHRR and THR of EP were reduced by 27.6% and 29.0%, respectively. In addition, the impact strength, tensile strength, and storage modulus increased by 65.0%, 29.0%, and 13.2%, respectively [17].

In this study, HCNP was synthesized through polymerization with phosphonitrilic chloride trimer (HCCP) and 4,4'-diaminobenzanilide (DABA). FR-PA66 was prepared by a twin-screw extrusion reaction with the amino groups of HCNP and end carboxyl groups of PA66. The structure of HCNP and the thermal stability, flame retardancy, and toxic properties of FR-PA66 were explored by FTIR, TGA, TG-IR, cone calorimetry, and SEM.

2. Results
2.1. Characterization of HCNP
2.1.1. FTIR Analysis

The FTIR spectra of HCNP are shown in Figure 1b. The transmittance peaks at 2970 and 2885 cm^{-1} of HCNP are assigned to the -NH_2 stretching bands of aniline; the peak at 2927 cm^{-1} shows the -NH bonds in the benzamide group; and the NH in-plane bending vibration and the coupling effect between δNH and $\nu C-N$ of HCNP are at 955 and 1390 cm^{-1}, respectively. The peak at 1612 cm^{-1} of HCNP indicates the presence of the C=O bonds of benzamide in the HCNP. The transmittance bands for P=N and P–N stretching vibration in HCNP are at 1197 and 1164 cm^{-1}, respectively. The band at

760 cm^{-1} of HCNP coincides with the CH stretching vibration of the para-substitution of a benzene ring.

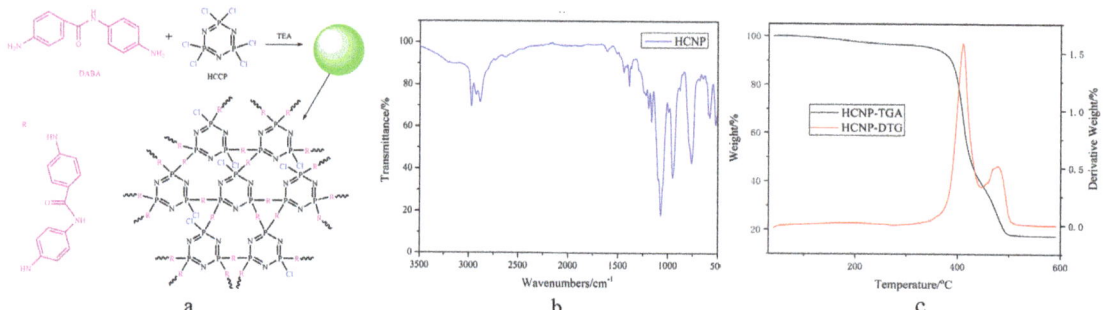

Figure 1. The synthetic routes and FTIR spectra of HCNP. (**a**) The synthetic routes of HCNP, (**b**) the FTIR spectra of HCNP, and (**c**) the TGA curves of HCNP.

2.1.2. Microtopography of HCNP

The microtopography of HCNP is shown in Figure 2. The SEM and TEM results clearly showed that the HCNP presents a spherical structure, and the diameter of the HCNP microsphere is about 2 um with a smooth surface. However, also seen are a few irregular edges on the surface of HCNP, as shown in Figure 2a,b, which are inferred to be incomplete reactions of DABA rich in amino groups according to the equation. Moreover, the TEM results (Figure 2d) showed that HCNP microspheres are amorphous.

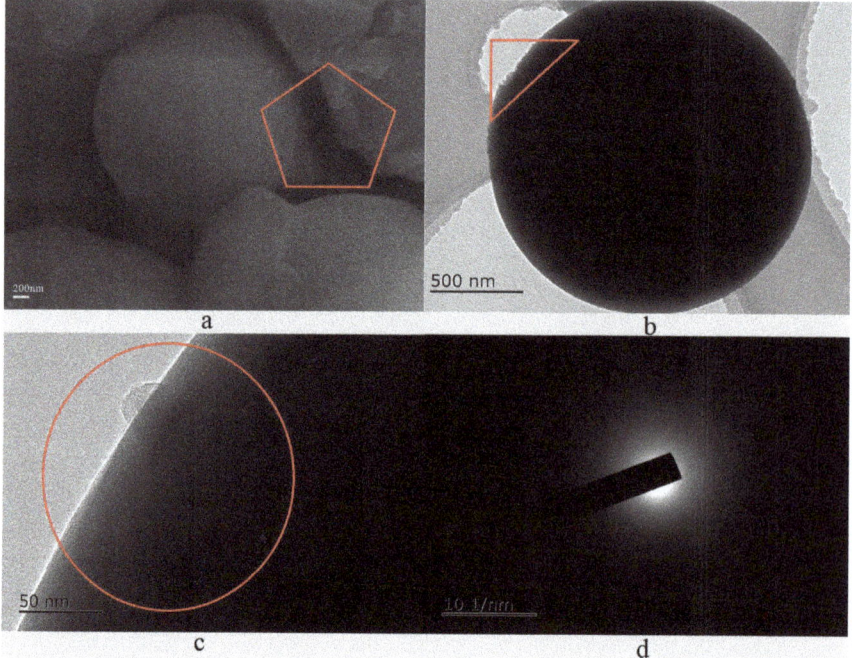

Figure 2. The microtopography of HCNP microsphere. (**a**) The SEM image of HCNP, (**b**–**d**) the TEM images of HCNP.

2.2. LOI and UL-94 Test Results

The LOI values and UL94 results with 1.6 mm thickness of FR-PA66 are listed in Table 1, indicating that suitable HCNP contents (9 wt%) induce a higher LOI value of 30%, and UL94 V-0 rating is achieved at the same time.

Table 1. The LOI and UL-94 results of FR-PA66.

Samples	PA66/wt%	HCNP/wt%	LOI	UL94
A0	100	0	24	V-2
A1	97	3	25	V-2
A2	95	5	26	V-2
A3	93	7	28.5	V-1
A4	91	9	30	V-0

2.3. Thermal Stability of FR-PA66

2.3.1. TGA Analysis

Figure 3a shows the thermal degradation curves of FR-PA66 under a nitrogen atmosphere at a heating rate of 10 °C/min. It is interesting to see that there are two sharp weight loss peaks of FR-PA66 on account of introducing HCNP, shown in Figure 3b.

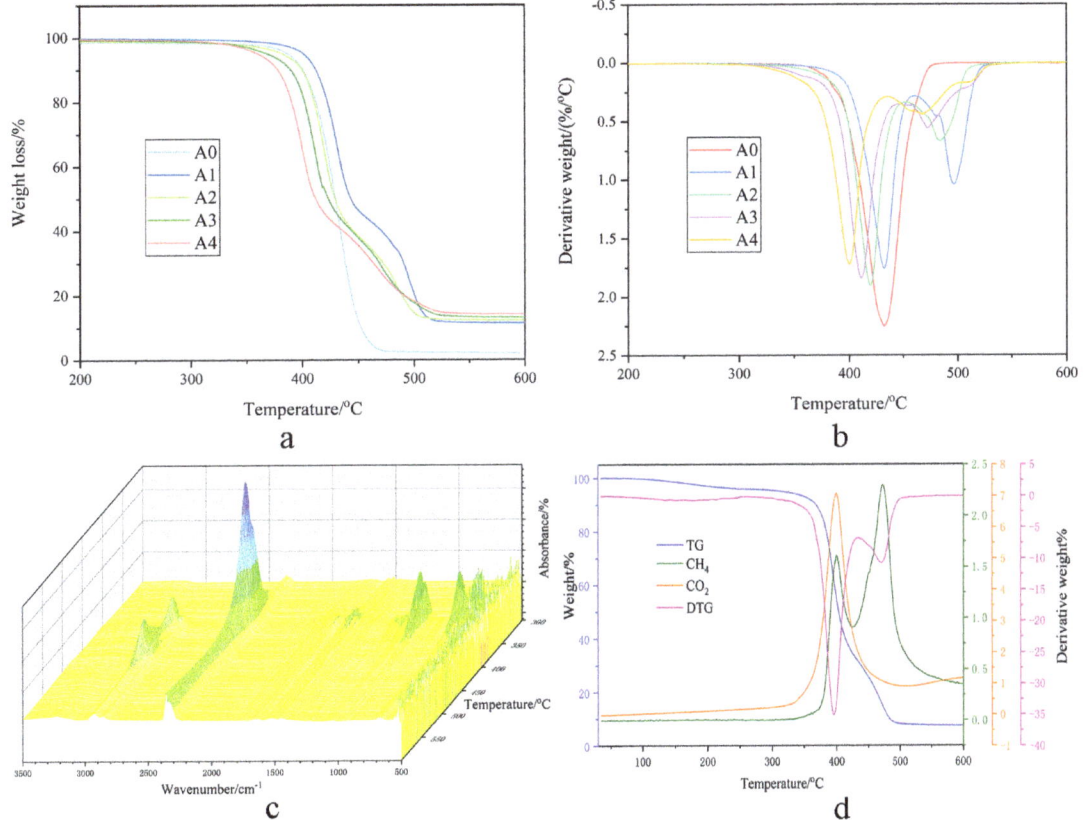

Figure 3. The thermal properties curves of FR–PA66. (**a**) The TGA curves of FR–PA66, (**b**) the DTG curves of FR–PA66, and (**c**,**d**) the TG–IR curves of FR–PA66.

The initial temperature (defined as 5% mass loss temperature, T_i) of the pristine PA66 is 387 °C, and the char residue only reaches 2.1 wt% at 600 °C. As for FR-PA66, the T_i and first decomposition peak (T_{max1}) decrease continuously with the increase in HCNP. When the HCNP content reaches 9 wt%, although the T_i and T_{max1} decrease from 387 °C and 432 °C to 360 °C and 400 °C, respectively, the char residues of FR-PA66 increase steadily from 39.65 wt% to 67.5 wt% at T_{max1}, and 2.10 wt% to 14.1 wt% at 600 °C.

The data details are noted in Table 2.

Table 2. The TGA and DTG detail of FR-PA66.

Samples	T_i/°C	T_{max1}/°C	T_{max2}/°C	Residue%/(T_{max1})	Residue% (Final)
A0	387	432	-	39.7	2.1
A1	401	432	497	62.8	11.4
A2	382	419	484	66.0	12.3
A3	368	411	473	67.4	13.2
A4	360	400	467	67.5	14.1

T_{max1} related to the first degradation of FR-PA66. T_{max2} related to the second degradation of FR-PA66.

2.3.2. TG-IR Analysis

Figure 3c displays the TG-IR curves of FR-PA66 under an inert atmosphere in the range of 300–600 °C. At 380 °C, the band at 669 cm^{-1} was assigned to the aldehyde such as acetaldehyde and aliphatic ketones, and another carbonyl band located at 1508 cm^{-1} was assigned to the amide carbonyl. At 400 °C, the absorbance peaks observed are as follows: 931 cm^{-1} (NO_2), 966 cm^{-1} (C=N), 1000–1200 cm^{-1} (P=O and C-P=O vibrations), 1350–1700 cm^{-1} (NH and C=O), 1706 and 1769 cm^{-1} (most probably the aliphatic ketone and cyclopentanone, the major degradation product of PA66), 2365 cm^{-1} (CO_2, confirmed in Figure 3d), and 1454 and 2929 cm^{-1} (the P-alkane) [18,19]. In addition, with gradually increasing temperature, at approximately 470 °C, the peak at 2860 cm^{-1} indicates CH_4 as the major volatile released (Figure 3d).

2.4. The Flame Retardancy of FR-PA66

Cone calorimetry is the most useful technique to evaluate the flame retardancy of materials. The heat release rate (HRR), THR, ML, mass loss rate (MLR), smoke production rate (SPR), and TSR profiles of FR-PA66 are shown in Figure 4. When the content of HCNP reaches 9 wt%, the pHRR declines significantly from 717.5 kW/m^2 (pristine PA66) to 373.7 kW/m^2, the THR reduces from 145.1 MJ/m^2 to 106.7 MJ/m^2 (Figure 4a), the ML remains at 7.5 wt% when the flame is extinguished, which is 7.0 wt% higher than that of the pristine PA66 (0.5 wt%) (Figure 4b), and the TSR decreases dramatically from 2667.0 m^2/m^2 of neat PA66 to 944.8 m^2/m^2 of FR-PA66 (Figure 4c).

The data details are noted in Table 3.

Table 3. The data of cone calorimetry.

Samples	pkHRR/(kW/m^2)	THR/(MJ/m^2)	ML/%	TSR/(m^2/m^2)
A0	717.5	145.1	99.5	2667.0
A4	373.7	106.7	92.5	944.8

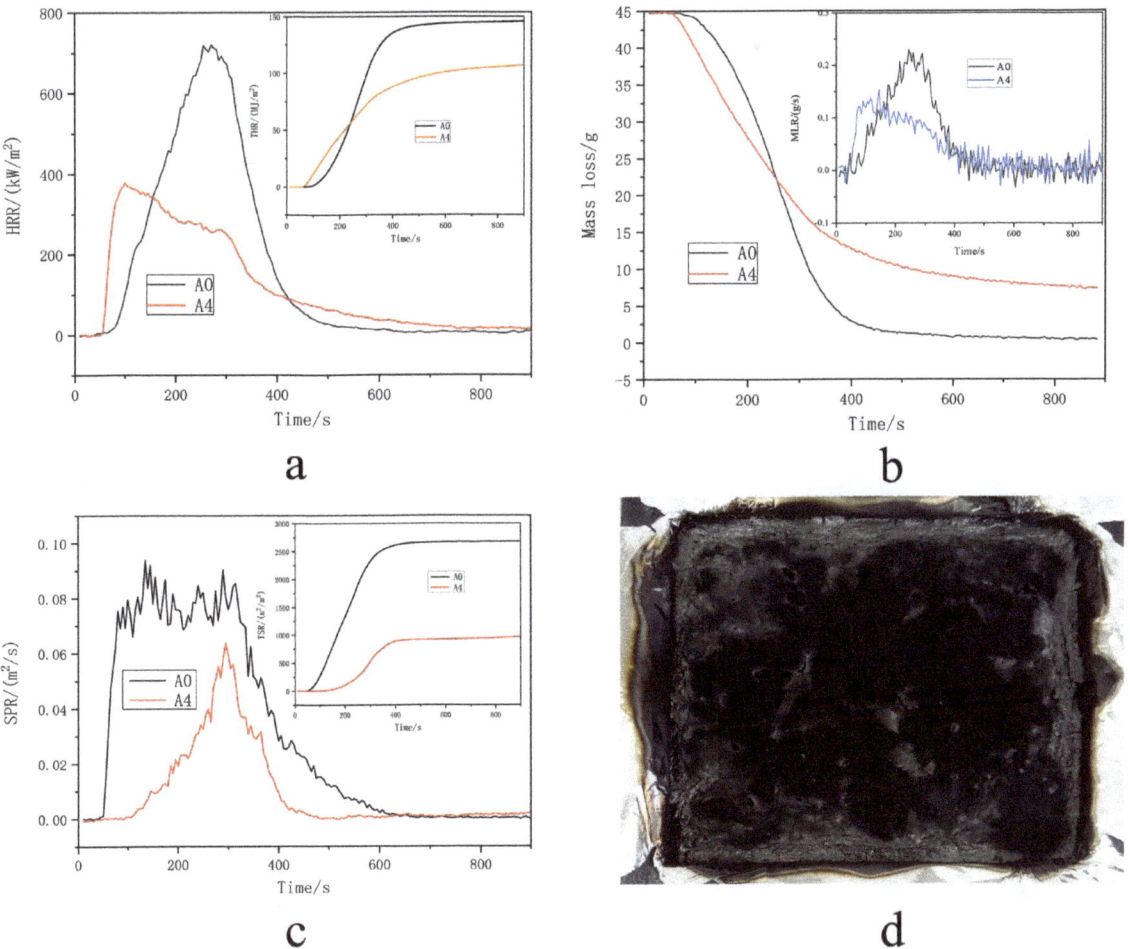

Figure 4. The cone curves of FR–PA66. (**a**) The HRR and THR curves of FR–PA66, (**b**) the MLR and ML curves of FR–PA66, (**c**) the SPR and TSR curves of FR–PA66, and (**d**) a picture of FR–PA66 after the cone test.

2.5. Microtopography

Figure 5 shows the SEM pictures of the residual char after the cone calorimetry test of FR-PA66, clearly indicating that the charred layer of FR-PA66 has a denser carbonized state (Figure 5b) than pure PA66 (Figure 5a). Figure 5c,d reveal that the char layer has a dense, high-surface-area, folded structure, and this particular morphology greatly blocks the heat transfer from the combustion of FR-PA66.

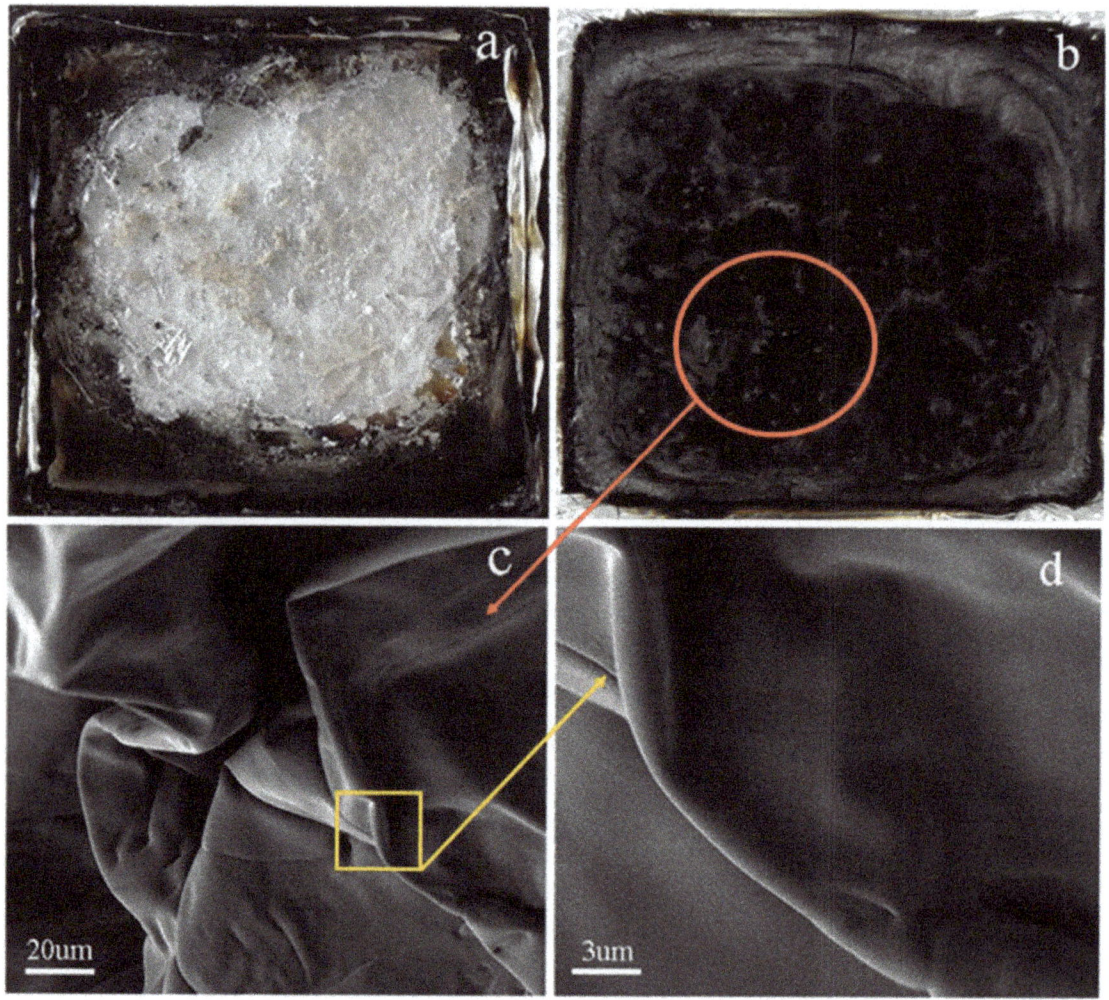

Figure 5. The pictures of residual char and SEM morphology. (**a**) the picture of pure PA66 after Cone calorimetry test; (**b**) the picture of FR–PA66 after Cone calorimetry test; (**c**,**d**) the SEM mophology images of FR–PA66; and red circle and yellow box: Zoom on the specified area.

3. Discussion

3.1. Thermal Stability of FR-PA66

Regarding the TGA results of FR-PA66, the T_{max1} gradually shifted to a lower temperature than that of the neat PA66 with increasing HCNP due to the low T_i of pure HCNP. Figure 1c shows that the T_i of HCNP is 352 °C because of the P-N bond priority breakage between HCCP and DABA under lower temperatures. It is interesting to see that the P-containing alkanes of FR-PA66 are detected by the TG-IR test under 400 °C, which is significantly different from the thermal decomposition products of pure PA66 [20]. Moreover, the R-P· radicals produced by HCNP can catalyze and accelerate the breaking of C-N bonds at the α position of the PA66 molecular chain, resulting in the formation of unstable amino radicals and carbonyl radicals, leading to the decomposition of the FR-PA66. Therefore, the T_i decreases significantly in the process of heating because the nitrogen oxide and vapor are easily released accompanying the thermal decomposition of nitrogen-containing compounds at low temperatures [21], which is confirmed by the TG-IR results. Moreover, it is worth noting that FR-PA66 shows two decomposition peaks compared with pure PA66, and the T_{max2} is higher than the T_{max1} of pure PA66 irrespective of the content of HCNP. This phenomenon may be due to the dehydration reaction of CNHO groups and the dehydrogenation reaction of α-CH$_2$ facilitated by the R-P· free radical substances produced from HCNP, accelerating the formation of the protective compact char layers [22].

3.2. Flame Retardancy of FR-PA66

The flame retardancy of PA66 obviously improved with the introduction of HCNP, and the significant decrease in the HRR, THR, ML, and TSR indicates the enhanced flame retardancy of FR-PA66, which is in keeping with the results of other studies on flame-retardant polyamides [23,24]. This phenomenon is attributed to the abundant release of phosphorous- and carbonyl-containing compounds in the initial heating stage. On the one hand, the R-P· groups (confirmed by the TG-IR results) obtained from HCNP under continuous heating can capture the free radicals produced from the degradation of PA66 and interrupt the chain reaction. On the other hand, the stable multilayer carbonaceous chars (Figure 5) produced from the reaction between HCNP and the decomposition products of PA66 can insulate the underlying substrate from the heat source and slow down both the heat and mass transfer [25]. Thus, it can be summarized that HCNP can effectively improve the flame retardancy of PA66 following the gas-phase flame retardant mechanism.

4. Materials and Methods

4.1. Materials

HCCP and DABA were provided by energy chemical Co., Ltd. (ShangHai, China). Triethylamine, acetonitrile, and anhydrous ethanol were purchased from Macklin Co., Ltd. (ShangHai, China). PA66 was supplied by Dupont China Co., Ltd. (ShangHai, China).

4.2. Synthesis of HCNP

We accurately weighed and dissolved 0.696 g HCCP and 1.362 g DABA in 200 mL acetonitrile; the mixed solution was poured into the three flasks with nitrogen as a protection gas placed in an ultrasonic cleaner under 40 kHz for dispersing at a temperature of 50 °C. Then, 3 mL TEA was added to the three flasks after sonication for 10 min, carrying on the reaction for 4–6 h. The products were washed 3–4 times with anhydrous ethanol and deionized water. Finally, the HCNP was obtained after 24 h at 80 °C in a vacuum drying oven.

4.3. Preparation of FR-PA66

PA66 pellets and HCNP powder were dried for 4 h at 100 °C prior to extrusion. The FR-PA66 was prepared by twin-screw extrusion reaction in the range of 260–270 °C. The test samples were manufactured using a TTI-95G injection molding machine under 265 °C with a 180 r/min rotation speed. Figure 6 reveals the microtopography of pure PA66, HCNP, and FR-PA66. It can be seen from Figure 6c that the HCNP microspheres were dispersed more uniformly without obvious phase interfaces, and the peeling phenomenon appeared on the surface of the microspheres (Figure 6d), indicating that the terminal carboxyl groups of PA66 were chemically bonded with the amino groups on the surface of the microspheres during the extrusion reaction, which destroyed the regularity of the surface of the microspheres (Figure 6b, smooth surface and no peeling phenomenon) and led to the core–shell separation of the microspheres (Figure 6d), confirming the occurrence of the in situ loading reaction.

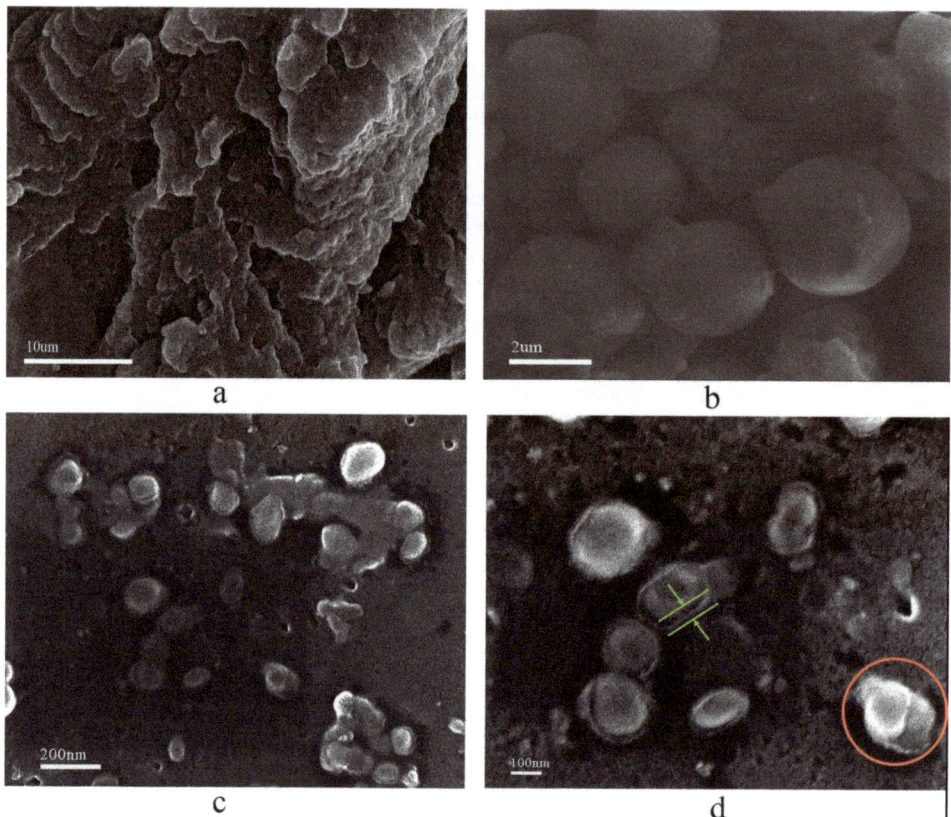

Figure 6. The microtopography of FR–PA66. (**a**) The SEM microtopography of pure PA66, (**b**) the SEM microtopography of HCNP only, and (**c**,**d**) the SEM microtopography of FR–PA66. And green lines and arrows and red circle: the peeling phenomenon on the surface of the microsphere.

4.4. Measurement Methods

FTIR spectra (wavelength range: 4000–500 cm^{-1}, resolution: 4.0 cm^{-1}) were recorded using a Nicolet iS10 FTIR spectrophotometer by using thin KBr pellets.

TGA was performed using a Q50 thermal gravimetric analyzer (TA instruments, New Castle, DE, America); 3–5 mg samples were heated from 50 to 600 °C at a heating rate of 10 °C/min under a nitrogen atmosphere.

The decomposition products were identified using a Q50 thermogravimetric analyzer coupled to a Nicolet Nexus spectrometer (TG-IR). The sample (approximately 8 mg) was heated in an open alumina crucible from 25 to 600 °C at a heating rate of 10 °C/min.

Cone calorimetry was measured by ENISO1716 (FTT, East Grinstead, England) with a $100 \times 100 \times 4$ mm^3 sample at 35 kW/m^2.

The microstructure study of the HCNP and char layers formed during the combustion in the cone calorimetry test was conducted using the scanning electron microscope ZEISS Sigma HDTM. Images were obtained under vacuum at a voltage of 5 kV.

5. Conclusions

The FR-PA66 was prepared by twin-screw extrusion reaction in situ loading HCNP. When the content of HCNP reached 9 wt%, the LOI and UL94 values of FR-PA66 improved up to 30% and V-0, respectively. Although the T_i and T_{max1} of FR-PA66 decreased by 27 °C and 32 °C, respectively, lower than that of the pristine PA66, the residue at T_{max1} and 600 °C was raised 27.9 wt% and 12.0 wt%, respectively, showing the improvement in the thermal stability of PA66. In addition, the TG-IR analysis showed the formation of nitrogen- and phosphorus-containing radicals under heating, which further confirmed the gas flame retardant mechanism of HCNP. Moreover, the pHRR, THR, and TSP of FR-PA66 decreased 47.9%, 26.5%, and 68.9%, respectively, accompanied by an increase in ML of about 7.0 wt%, revealing that the flame retardancy of PA66 had been modified markedly by adding HCNP. The SEM microtopography images also showed excellent compatibility and dispersion of HCNP in the PA66 matrix.

Author Contributions: Conceptualization, W.L.; validation, C.L.; investigation, J.L. and C.Z.; resources, Y.Z.; data curation, Y.C. and L.Z.; writing—original draft preparation, L.W.; writing—review and editing, W.L.; supervision, C.L.; project administration, W.L.; funding acquisition, W.L. All authors have read and agreed to the published version of the manuscript.

Funding: This study was supported by the youth project of science and technology research program of Chongqing Education Commission of China (KJQN202001119, KJQN202201130) and the Natural Science Foundation of Chongqing Bureau of Science and Technology (CSTB2022NSCQ-MSX1052).

Institutional Review Board Statement: Not applicable.

Informed Consent Statement: Not applicable.

Data Availability Statement: Data are contained within the article.

Conflicts of Interest: The authors declare no conflict of interest.

References

1. Zhang, H.; Lu, J.L.; Yang, H.Y.; Lang, J.Y.; Yang, H. Comparative Study on the Flame-Retardant Properties and Mechanical Properties of PA66 with Different Dicyclohexyl Hypophosphite Acid Metal Salts. *Polymers* **2019**, *11*, 1956. [CrossRef] [PubMed]
2. Li, H.; Wang, X.H.; Sun, J.J.; Yang, C.B.; Wu, J.; Ying, J.G.; Liu, P. Preparation and Technology of Flame Retardant Reinforced Toughening Nylon 66 Composites. *J. Eng. Plas. Appl.* **2012**, *40*, 49–53.
3. Zhan, Z.S.; Shi, J.S.; Zhang, Y.; Zhang, Y.L.; Zhang, B.; Liu, W.B. The study on flame retardancy synergetic mechanism of magnesium oxide for PA66/AlPi composite. *Mater. Res. Express* **2019**, *6*, 115317. [CrossRef]
4. Zhan, Z.S.; Li, B.; Xu, M.J.; Guo, Z.H. Synergistic effects of nano-silica on aluminum diethylphosphinate/polyamide 66 system for fire retardancy. *High. Perform. Polym.* **2016**, *28*, 140–146. [CrossRef]
5. Luo, D.; Duan, W.F.; Liu, Y.; Chen, N.; Wang, Q. Melamine cyanurate surface treated by nylon of low molecular weight to prepare flame-retardant polyamide 66 with high flowability. *Fire Mater.* **2019**, *43*, 323–331. [CrossRef]
6. Guo, Z.; Bao, M.; Ni, X.Y. The synthesis of meltable and highly thermostabletriazine-DOPO flame retardant and its application in PA66. *Polym. Advan. Technol.* **2021**, *32*, 815–828. [CrossRef]
7. Weil, E.D.; Levchik, S.V. Current Practice and Recent Commercial Developments in Flame retardant of Polyamides. *J. Fire Sci.* **2004**, *22*, 251–264. [CrossRef]
8. Levchik, S.V.; Levchik, G.F.; Balabanovich, A.I.; Weil, E.D.; Klatt, M. Phosphorus oxynitride: A thermally stable fire retardant additive for polyamide 6 and poly(butylene terephthalate). *Macromol. Mater. Eng.* **1999**, *264*, 48–55. [CrossRef]
9. Jiang, P.; Gu, X.Y.; Zhang, S.; Wu, S.D.; Zhao, Q.; Hu, Z.W. Synthesis, Characterization, and Utilization of a Novel Phosphorus/Nitrogen-Containing Flame retardant. *Ind. Eng. Chem. Res.* **2015**, *54*, 2974–2982. [CrossRef]
10. Ding, Y.; Swann, D.J.; Sun, Q.; Stoliarov, I.S.; Kraemer, H.R. Development of a pyrolysis model for glass fiber reinforced polyamide 66 blended with red phosphorus: Relationship between flammability behavior and material composition. *Compos. Part B-Eng.* **2019**, *176*, 107263. [CrossRef]
11. Sui, Y.L.; Sim, H.F.; Shao, W.D.; Zhang, C.L. Novel bioderived cross-linked polyphosphazene microspheres decorated with FeCo-layered double hydroxide as an all-in-one intumescent flame retardant for epoxy resin. *Compos. Part B-Eng.* **2022**, *229*, 109463. [CrossRef]
12. Zhu, Y.Z.; Wu, W.; Xu, T.; Xu, H.; Zhong, Y.; Zhang, L.P.; Ma, Y.M.; Sui, X.F.; Wang, B.J.; Feng, X.L.; et al. Preparation and characterization of polyphosphazene-based flame retardants with different functional groups. *Polym. Degrad. Stabil.* **2022**, *196*, 109815. [CrossRef]
13. Zhou, X.; Qiu, S.L.; Mu, X.W.; Zhou, M.T.; Cai, W.; Song, L.; Xing, W.Y.; Hu, Y. Polyphosphazenes-based flame retardants: A review. *Compos. Part B-Eng.* **2020**, *202*, 108397. [CrossRef]
14. Zhou, X.; Qiu, S.L.; He, L.X.; Cai, W.; Chu, F.K.; Zhu, Y.L.; Jiang, X.; Song, L.; Hu, Y. Bifunctional linear polyphosphazene decorated by allyl groups: Synthesis and application as efficient flame-retardant and toughening agent of bismaleimide. *Compos. Part B-Eng.* **2022**, *233*, 109653. [CrossRef]

15. Qiu, S.L.; Ma, C.; Wang, X.; Zhou, X.; Feng, X.M.; Yuen, R.K.K.; Hu, Y. Melamine-containing polyphosphazene wrapped ammonium polyphosphate: A novel multifunctional organic-inorganic hybrid flame retardant. *J. Hazard. Mater.* **2018**, *344*, 839–848. [CrossRef]
16. Qiu, S.L.; Hu, Y.X.; Shi, Y.Q.; Hou, Y.B.; Kan, Y.C.; Chu, F.K.; Sheng, H.B.; Yuen, R.K.K.; Xing, W.Y. In situ growth of polyphosphazene particles on molybdenum disulfide nanosheets for flame retardant and friction application. *Compos. Part A-Appl. Sci. Manuf.* **2018**, *114*, 407–417. [CrossRef]
17. Zhou, M.T.; Zeng, D.S.; Qiu, S.L.; Zhou, X.; Cheng, L.; Xu, Z.M.; Xing, W.Y.; Hu, Y. Amino-functionalized carbon nanotubes/polyphosphazene hybrids for improving the fire safety and mechanical properties of epoxy. *Prog. Nat. Sci.-Mater. Int.* **2022**, *32*, 179–189. [CrossRef]
18. Dong, L.P.; Huang, S.C.; Li, Y.M.; Deng, C.; Wang, Y.Z. A Novel Linear-Chain Polyamide Charring Agent for the Fire Safety of Noncharring Polyolefin. *Ind. Eng. Chem. Res.* **2016**, *55*, 7132–7141. [CrossRef]
19. Jiang, P.; Zhao, Q.; Zhang, S.; Gu, X.Y.; Hu, Z.W.; Xu, G.Z. Flammability and Char Formation of Polyamide 66 Fabric: Chemical Grafting versus Pad-Dry Process. *Ind. Eng. Chem. Res.* **2015**, *54*, 6085–6092. [CrossRef]
20. Schartel, B.; Kunze, R.; Neubert, D. Red phosphorus-controlled decomposition for fire retardant PA 66. *J. Appl. Polym. Sci.* **2002**, *83*, 2060–2071. [CrossRef]
21. Guan, X.Y.; Zheng, G.Q.; Dai, K.; Liu, C.; Yan, X.R.; Shen, C.Y.; Guo, Z.H. Carbon Nanotubes-Adsorbed Electrospun PA66 Nanofiber Bundles with Improved Conductivity and Robust Flexibility. *ACS Appl. Mater. Inter.* **2016**, *8*, 14150–14159. [CrossRef] [PubMed]
22. Xu, G.; Zhang, B.; Xing, J.; Sun, M.M.; Zhang, X.G.; Li, J.H.; Wang, L.; Liu, C.Z. A facile approach to synthesize in situ functionalized graphene oxide/epoxy resin nanocomposites: Mechanical and thermal properties. *J. Mater. Sci.* **2019**, *54*, 13973–13989. [CrossRef]
23. Li, L.; Liu, X.L.; Shao, X.M.; Jiang, L.; Huang, K.; Zhao, S. Synergistic effects of a highly effective intumescent flame retardant based on tannic acid functionalized graphene on the flame retardant and smoke suppression properties of natural rubber. *Compos. Part A-Appl. Sci. Manuf.* **2020**, *129*, 105715. [CrossRef]
24. Wang, G.; Huang, Y.; Hu, X. Synthesis of a novel phosphorus-containing polymer and its application in amino intumescent fire resistant coating. *Prog. Orga. Coat.* **2013**, *76*, 188–193. [CrossRef]
25. Liu, Y.; Gao, Y.S.; Zhang, Z.; Wang, Q. Preparation of ammonium polyphosphate and dye co-intercalated LDH/polypropylene composites with enhanced flame retardant and UV resistance properties. *Chemosphere* **2021**, *277*, 130370. [CrossRef]

Disclaimer/Publisher's Note: The statements, opinions and data contained in all publications are solely those of the individual author(s) and contributor(s) and not of MDPI and/or the editor(s). MDPI and/or the editor(s) disclaim responsibility for any injury to people or property resulting from any ideas, methods, instructions or products referred to in the content.

Article

Mechanical Characterization of Hybrid Nano-Filled Glass/Epoxy Composites

Ali A. Rajhi

Department of Mechanical Engineering, College of Engineering, King Khalid University, Abha 61421, Saudi Arabia; arajhi@kku.edu.sa

Abstract: Fiber-reinforced polymer (FRP) composite materials are very versatile in use because of their high specific stiffness and high specific strength characteristics. The main limitation of this material is its brittle nature (mainly due to the low stiffness and low fracture toughness of resin) that leads to reduced properties that are matrix dominated, including impact strength, compressive strength, in-plane shear, fracture toughness, and interlaminar strength. One method of overcoming these limitations is using nanoparticles as fillers in an FRP composite. Thereby, this present paper is focused on studying the effect of nanofillers added to glass/epoxy composite materials on mechanical behavior. Multiwall carbon nanotubes (MWCNTs), nano-silica (NS), and nano-iron oxide (NFe) are the nanofillers selected, as they can react with the resin system in the present-case epoxy to contribute a significant improvement to the polymer cross-linking web. Glass/epoxy composites are made with four layers of unidirectional E-glass fiber modified by nanoparticles with four different weight percentages (0.1%, 0.2%, 0.5%, and 1.0%). For reference, a sample without nanoparticles was made. The mechanical characterizations of these samples were completed under tensile, compressive, flexural, and impact loading. To understand the failure mechanism, an SEM analysis was also completed on the fractured surface.

Keywords: nanocomposites; glass fiber-reinforced polymer (GFRP); multiwall carbon nanotube (MWCNT); nano-silica (NS); nano-iron oxide (NFe); mechanical characterizations

1. Introduction

Because of their superior stiffness, strength, low density, light weight, resistance to corrosion, superior electrical properties, and ease of manufacture, fiber-reinforced polymer nanocomposites stand out from all other materials on the market and have attracted notable studies [1–3]. Because of their exceptional qualities, they have been widely used in a variety of applications, including the construction of buildings and sports equipment as well as in the automotive, aviation, and defense industries [4,5]. Each application, however, is subject to a unique set of circumstances, including cyclic loading, impact conditions, stretching conditions, and deformation characteristics. Investigating the behavior of these kinds of composite materials under various conditions is crucial.

Chang [6] studied the effect of carbon-fiber-reinforced composites and GFRP with the addition of MWCNTs. It was reported that the tensile strength improved by 34.7% with the addition of the MWCNTs, and the flexural strength improved by 22.16%. Markand et al. [7] investigated the effects of adding carbon nanotube (CNT) to GFRP composite laminates when subjected to interlaminar shear and flexural loading using different percentages of hardened resin (phr) (0.25, 0.5, 0.75, and 1 phr). They discovered that CNT at 0.75 phr has the best mechanical properties, improving the flexural strength and interlaminar shear strength (ILSS) by 15.7% and 9.2%, respectively.

In an experimental investigation, M.R. Ayatollahi et al. [8] investigated the effects of the MWCNT aspect ratio on the electrical and mechanical properties of epoxy/MWCNT composite plates. They established that the aspect ratio has a significant impact on the electrical and mechanical capabilities of nanocomposite materials, with smaller MWCNTs exhibiting significantly superior qualities. Laminated fiberglass/epoxy composites were examined for their mechanical, vibrational, and damping properties by M Rafiee et al. [9,10], using a variety of carbon nanofillers, such as multiwall carbon nanotubes, graphene oxide, reduced graphene oxide, and graphene nanoplatelets. According to the experimental findings, as the nanoloading increased, the damped natural frequencies and tensile characteristics of the nanocomposites also increased.

Mostovoy et al. [11] studied basalt-fiber-reinforced epoxy composites modified with nano-graphene oxide. The nano-graphene oxide was functionalized with aminoacetic acid and APTES, which enabled the functionalization. The samples were prepared with different weight percentages of the nano-graphene oxide, and various physio-mechanical tests were conducted. The results show that the 0.5 wt% graphene oxide samples have a better tensile strength (1830 MPa) when compared to the neat composite (160 MPa). The tensile modulus improved by 31% and 19% for the modified composite, whereas the flexural strength improved by only 9% and 13%.

Josh et al. [12] used carbon fiber as a reinforcement and epoxy as a matrix with nano-graphite oxide as the filler. Vacuum-assisted resin infusion molding was used to manufacture the composites. To add the nanofiller to the composite materials, two methods were used: (i) direct spraying onto carbon fiber with ethanol as a base, and (ii) mixing the nanoparticle in resin and using it to manufacture the composite. The first method of spraying is better, as the mixing of the nanoparticles included in the resin increases the density of the resin, which reduces the wettability of the fiber and may lead to more defects. They reported an improvement in transverse tensile strength of 8% and an improvement in interlaminar shear strength of 15% with the addition of the nanoparticles. However, not much improvement was observed along the longitudinal direction.

Bekeshev et al. [13] used the mineral filler ocher with epoxy resin to develop the composite material. Ocher with a size < 40 μm and these particles were first mechanically stirred and then sonicated to obtain a uniform distribution. Different samples were fabricated with different parts according to the mass of the ocher. The results show an outstanding improvement in tensile strength of 75%, a 20% improvement in tensile modulus, and an improvement in impact strength of 83%, whereas the flexural strength and flexural modulus improved by 30% and 58%, respectively. These products also led to an increase in the yield of carbonized structures from 54% to 58–76%, which led to the low flammability of the epoxy.

Investigations were completed by Jamali et al. [14] on how Graphene Oxide Nanoplatelet (GNOP) modification and Silica-GONP loading affected the mechanical characteristics of the basalt/epoxy composite. By using FTIR, STA, and Raman spectroscopy, the introduction of the silane organic chains on the surface of the GONPs was assessed. The specimens' mechanical strength reached their maximum levels at 0.4 weight percent S-GONPs. These specimens' tensile, flexural, and compressive strengths were higher than those of the basal/epoxy composite by 16%, 47%, and 51%, respectively. Additionally, the mechanical moduli improved. Further, it was discovered that silane alteration of the GONPs significantly changed the specimens' mechanical characteristics. Microscopic examinations revealed that the specimens filled with nanofiller had an improved interfacial adhesion between the basalt and the matrix.

Tian et al. [15] used a matrix made of sol–gel silica/epoxy nanocomposites to enhance the interfacial characteristics between the polymers and the fibers. The interfacial adhesion was significantly increased by the silica nanoparticles, as demonstrated in both micro- and macro-mechanical studies. When compared to the carbon fiber/epoxy system, the IFSS and transverse fiber bundle tension (TFBT) strength of the carbon fiber/20 wt% nanosilica-epoxy system rose by roughly 38% and 59%, respectively. For the CF/10 wt% nanosilica-epoxy system, the ILSS of the unidirectional laminar also rose by up to 13%. These gains can be attributed to the nanoparticle-enhanced, toughened matrix, which improves stress transfer and resists debonding by reducing stress concentration and dissipating more deformation energy.

Shu-quan et al. [16] studied the tensile behavior of CNT-modified epoxy composites which included tensile modulus and tensile strength. A proportional improvement in tensile properties (strength and modulus) was reported, with an increase in filler quantity up to 1.75% of the mass fraction. The optimum mass fraction obtained was 0.75% epoxy resin. If the mass fraction increases by more than 1.75%, then the tensile strength and tensile modulus are lower than the neat resin. Similarly, Zhou et al. [17] reported an improvement in the resistance to crack propagation upon induction of the CNT in the epoxy resin nanocomposite. However, Wong et al. [18] reported that the increase in the weight percentage of multiwall carbon nanotubes (MWCNTs) in polystyrene resin has adverse effects on tensile strength, tensile modulus, and failure strain. Thus, an optimum weight percentage of nanoparticles to be added into the resin system needs to be determined, which would enhance the mechanical properties of the composite materials. Improvements in the mechanical, rheological, thermal, and adhesion properties of the nano-modified polyester with nano-silica were observed because of the formation of a hydrogen bond between the silanol groups and the ester carbonyl group on the nano-silica surface in soft segments [19–22]. According to Sudirman et al., adding nano-silica to polyester resin improves the chain mobility of the polymer, which results in an improved order compared to pure resin [23].

Zheng et al. [24] carried out a similar study in which NS was used to modify epoxy resin, and this modified resin was used as a matrix system in composites with glass fiber as reinforcement. Three different weight percentages of NS—1%, 5%, and 7%—were used for the preparation of the different laminates. Under tensile loading, the tensile strength and the tensile modulus improved by 24% and 22%, respectively, and a relatively smaller improvement was reported for the compressive and shear strengths, which improved by 13% and 14%, respectively. Under bending, the strength also increased by 22%. The reason for this improvement was given as a strong covalent bond between nano-silica and the fiber surface, which led to a better transfer of load and stress from the fiber to the matrix, and vice versa. As the weight percentage of nano-silica was high, the uniform dispersion of NS in the resin was not possible, thereby the compressive and shear properties did not improve to that extent.

Thus, the addition of nanoparticles in composite materials has both encouraging and destructive effects on the latter's properties, which include both physical and mechanical properties. This change in the properties of a nanoscale hybrid composite depends on (i) the geometry of the nanofiller, (ii) the type of nanofiller, (iii) the type of resin system, (iv) the filler percentage, (v) the dispersion of nanoparticles in the resin system, and (vi) the manufacturing method. The present work aims to optimize the types of nanofillers and their quantities to improve mechanical properties. Three different nanoparticles were used—nano-iron oxide, nano-silica, and MWCNTs—with four different weight percentages of 0.1, 0.2, 0.5, and 1.0 wt%. Four different mechanical tests—tensile, compressive, bending, and impact—were performed. Epoxy resin was used as a matrix material and unidirectional E-glass fibers were used as reinforcement.

2. Material and Methods

2.1. Materials

As mentioned, Lapox L12 is an epoxy resin that is commercially available along with a K-6 hardener and was used in the preparation of the laminates due to its wide utilization in the industry. The chemical name of Lapox L12 is Diglycidyl Ether of Bisphenol. Table 1 displays the various properties of the Lapox L12 used in this work. Three different nanoparticles, as mentioned earlier, were supplied by Intelligent Materials Pvt. Ltd., Punjab, India. and were used as nanofillers. Later, these particles were subjected to SEM (Zeiss GeminiSEM 360, ZEISS Microscopy, Jena, Germany) analysis for quality assurance and characterization, which are discussed in detail in Section 3. Figure 1 shows the SEM images of these nanoparticles, and with this analysis, the size and the other specifications presented in Table 1 were determined. A 600 GSM E-glass fiber roll containing 34% chopped strand fiber and 66% unidirectional fiber was used as reinforcement.

Table 1. Properties of Lapox L12.

S. No	Properties	Value
1.	Appearance	Clear, viscous liquid
2.	Viscosity at 25 °C	9000–12,000 m Pas
3.	Specific gravity at 25 °C	1.1–1.2
4.	Solubility	30 g/25 mL
5.	Melting point	88–92 °C
6.	Pot life	6 h–8 h at 20 °C 5 h–7 h at 30 °C 3 h–5 h at 40 °C
7.	Tensile strength	70–80 MPa
8.	Elastic modulus in tension	4.0–4.8 GPa
9.	Glass transition temperature (DSC)	150–160 °C
10.	Co-efficient of linear thermal expansion	$45–55 \times 10^{-6}$ K^{-1}

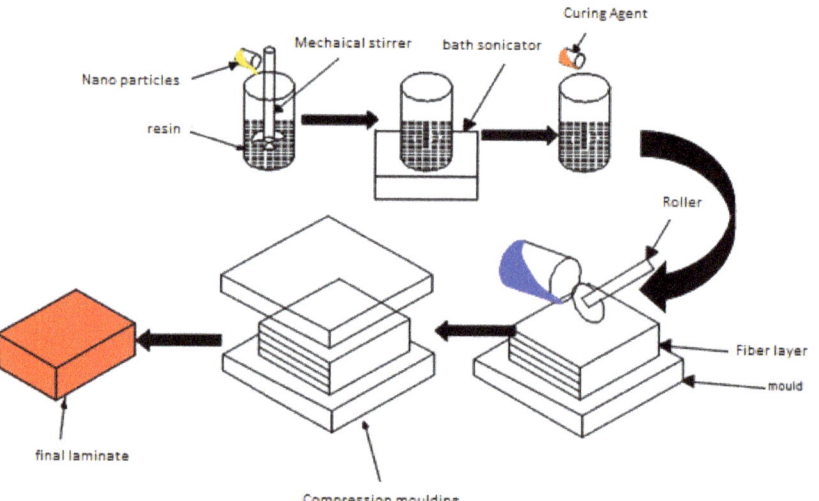

Figure 1. Schematic view of the fabrication of modified nano-polymer composites.

Figure 1 represents the schematic diagram of the fabrication of the modified fiber-reinforced polymer nanocomposites. Nanofillers were dispersed in a resin system in two stages, the first including mechanical stirring for an hour followed by bath sonication for an hour. Later, this modified resin system was used in the preparation of the samples using a hand lay-up technique followed by a compression molding technique at a pressure of 20 MPa. The samples required for the tensile, compression, bending, and impact tests were cut according to the ASTM (D3039, 695, D790 & D265) standard from a single laminate with a size of 250 × 250 mm^2. The same procedure was opted for the preparation of the different laminates from the different nanoparticles according to size and concentration, type of nanoparticle, and weight concentration, as tabulated in Table 2.

Table 2. Different compositions used in sample preparation.

S. No	Type of Resin	Type of Nanoparticle	Weight Percentage of Nanoparticles
1	EPOXY	MWCNT	0.10%
2	EPOXY	NS	0.10%
3	EPOXY	NFe	0.10%
4	EPOXY	MWCNT	0.20%
5	EPOXY	NS	0.20%
6	EPOXY	NFe	0.20%
7	EPOXY	MWCNT	0.50%
8	EPOXY	NS	0.50%
9	EPOXY	NFe	0.50%
10	EPOXY	MWCNT	1.00%
11	EPOXY	NS	1.00%
12	EPOXY	NFe	1.00%
13	EPOXY	-	-

2.2. Mechanical Characterizations

The mechanical properties of the samples were determined under tensile, compression, bending, and impact loading. The tensile tests were performed per the ASTM D3039 standard with a crosshead speed of 1 mm/min on the universal testing machine (UTM), Deepak Poly Plast, with a range of 5 tons. The specimens were cut from a single laminate with a width of 20 mm, a thickness of 2.5 mm, and a length of 240 mm. The gauge length of the samples was 100 mm with a clamping length of 50 mm. Three specimens from each sample were tested.

A static compression test was conducted on all samples using the UTM with a modified end-loading fixture as proposed by Shimokawa per the ASTM standard 695. The specimen was tightly calmed in the fixture with binding strips. The fixture has a pair of supporting guides that prevent the out-of-plane directional movement of the specimen under loading. The binding strips were used to apply the compression load at the end of the sections of the specimens.

Flexural tests under 3-point bending were also completed using a UTM per ASTM D790 to calculate the flexural properties (strength and modulus). The crosshead of the UTM moved with a constant speed of 2.0 mm per min and with a depth-to-span ratio of 1:16. The slope of the load-displacement graph was used to calculate the flexural modulus. The flexural strength of the specimen is the highest stress at failure on the tensile side. An average of three specimens were taken.

The Izod impact testing method was used to determine the energy observed (impact strength) for the breaking of the fabricated composite specimens per ASTM D256. Digital impact testing was used, which has advantages like more versatility, ease of operation, and the display of information with a high resolution. The machine has a maximum pendulum capacity of 25 J, a drop height of 0.61 m, and an impact velocity of 3.46 m/s. The specimen used was 8 cm × 3 cm × 0.4 cm. The specimens were cut along the longitudinal directions only.

3. Results and Discussion

3.1. SEM Analysis of the Nanoparticles

An SEM analysis of the nanoparticles was completed using a ZEISS GeminiSEM 360 machine. The images are shown in Figure 2 and the details of the results are tabulated in Table 3.

Figure 2. SEM images of (**a**) Nano-iron oxide, (**b**), Nano-silica, and (**c**) MWCNTs.

Table 3. Properties of nanoparticles.

Material Properties	Nano-Silica	Nano-Iron Oxide	MWCNT
Size (nm)	50	50	30–50
Shape	Spherical	Spherical	Cylindrical
Purity	99.9%	99.9%	99.8%
Color	White	Reddish	Black

3.2. Tensile Results

Figure 3 represents the load-displacement graphs under tensile loading of the neat glass-fiber-reinforced polymer (GFRP) and fiber-reinforced polymer nanocomposite GFRP with the different nanoparticles (NFe, MWCNT, NS) under different weight concentrations. From Figure 3a, it is observed that the breaking point of the fiber-reinforced nanocomposite with NFe as the filler under tensile load was earlier than that of the neat GFRP, except for the 0.1 wt% NFe sample, and this is because of the increase in the brittleness of the composite materials. The area under the curve of load-displacement is an indication of the modulus of toughness, which was calculated for all samples, and it was observed that the sample with 0.1 wt% NFe was 84.6% higher when compared to the neat GFRP. Whereas, for the other sample, this value is reduced to 29.4% when compared with the neat GFRP. It can also be observed that, as the weight percentage increased, the breaking point decreased. However, for the 0.1 wt% nano-iron oxide particles, there was an extended breaking point, and there was an increase in toughness of 84.6%, which was calculated as the area under the curve. There was a reduction in toughness of 29.4% for the 0.5 wt% nano-iron oxide fiber-reinforced polymer nanocomposite sample. Regarding the breaking point of the samples, in Figure 3b, it is observed that it is the same for all samples except for the 0.1 wt% MWCNT sample. The toughness improved by 14.6%, 30.82%, 52.66%, and 78.06% for the 0.1, 0.2, 0.5, and 1.0 wt% samples, respectively. Figure 3c shows that the breaking point extended by 20% for the 0.1 wt% nano-silica fiber-reinforced polymer nanocomposite sample, and thereafter it decreased for the rest of the samples. Breaking occurred 40% earlier when compared to the neat GFRP composite material. There were improvements in toughness of 43%, 83%, 63%, and 38.1% for the 0.1, 0.2, 0.5, and 1.0 wt% samples, respectively, of the fiber-reinforced polymer nanocomposites with nano-silica as the filler. From these graphs, the tensile strengths of the specimens were obtained, and they are compiled in Figure 4a–c. The common behavior of an increase in ultimate tensile strength for the 0.1 wt% sample was observed in all the modified GFRPs. In the case of the sample modified with MWCNTs, a slight decrement in ultimate tensile strength was observed for the 0.2 wt% sample, and thereafter it remained constant. For the sample of 0.1 wt%, a tensile strength improvement of 30% overall was seen. In the case of the samples modified with nano-iron oxide particles, the maximum tensile strength of the 0.2 wt%

sample exhibited an increment of 20% when compared with the neat GFRP. For the third set of samples modified with nano-silica, the maximum tensile strength for the 0.5 wt% sample exhibited an increment of 35% when compared with the normal GFRP. The good dispersion and exfoliation of the nanoparticles are the reasons for this improvement in tensile strength. The presence of an agglomerate caused a drop in the tensile strength of the specimens with a higher wt% of nanoparticles. This causes stress concentration rather than load transfer, and failure starts at this point.

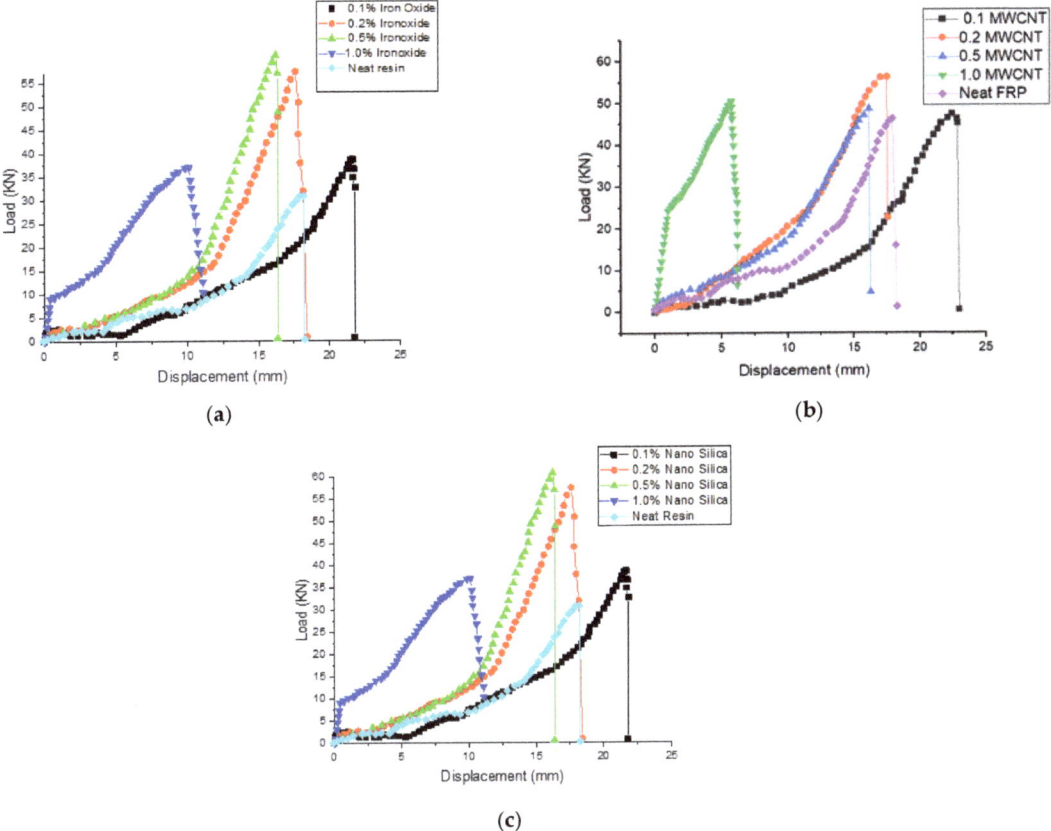

Figure 3. Load-displacement graphs of tensile tests for modified GFRP with (**a**) NFe, (**b**) MWCNTs, and (**c**) NS.

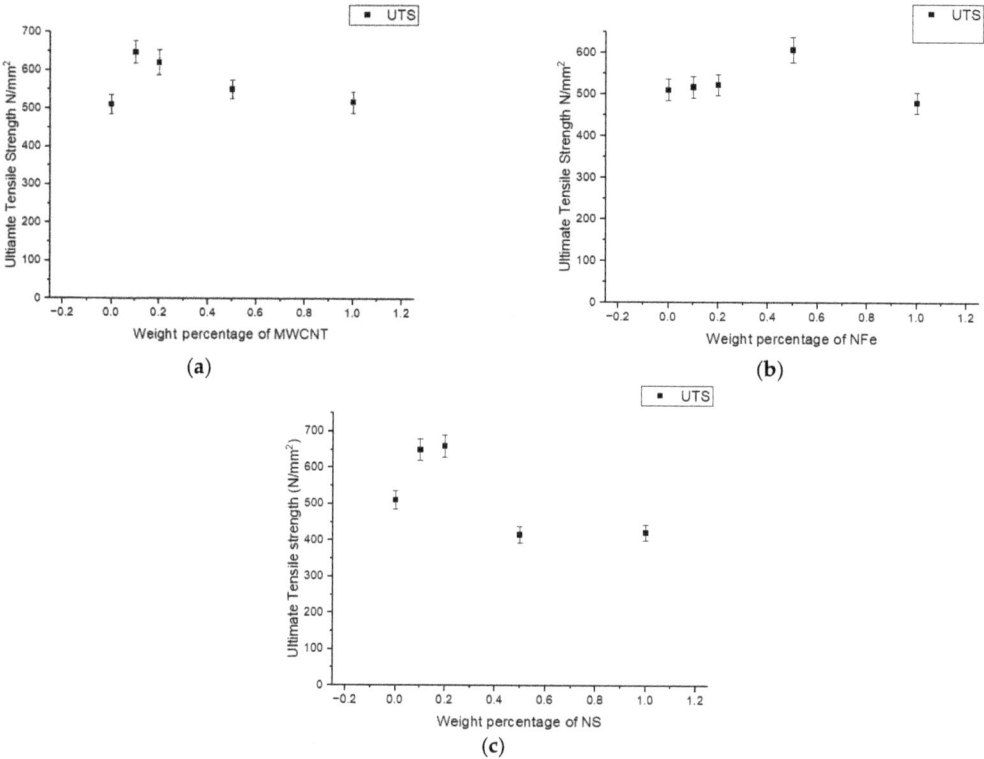

Figure 4. Variation in ultimate tensile strength of modified GFRP samples with different weight percentages of (**a**) MWCNTs, (**b**) NFe, and (**c**) NS.

3.3. Compression Results

As described in the above section, the compression test was carried out per ASTM D695, with a sample size of 25 × 25 mm². The ultimate compressive strengths for the different weight percentages of the nanoparticle-fiber-reinforced nanocomposites are plotted in Figure 5a–c. The inclusion of the MWCNTs increased the ultimate compressive strength (UCS) for the MWCNT-set fiber-reinforced polymer nanocomposites, as seen in Figure 5a. For the 0.1 weight percent MWCNT sample, the UCS was 1.27 times higher than the clean conventional composite, and for the 0.2 weight percent MWCNT sample, the UCS increased by 1.36 times. For the 0.5 wt% and 1.0 wt% MWCNT samples, the percentage increase was almost the same, which was 26% when compared with the neat conventional composite. A proportionate increase in the UCS for the first two sets of samples was due to improvements in the resin properties with the presence of MWCNTs. As in compression, the load was mostly taken up by the matrix material, and with the addition of MWCNTs, the resin properties increased, thereby the UCS also improved up to the 0.2 wt% sample. However, with further increases in the weight percentage of MWCNTs in the fiber-reinforced polymer of 0.5 wt% and 1.0 wt%, a reduction in the UCS was observed, as a higher wt% of MWCNTs led to a decrease in the dispersion of the nanoparticles due to an increase in the van der Waals force between them.

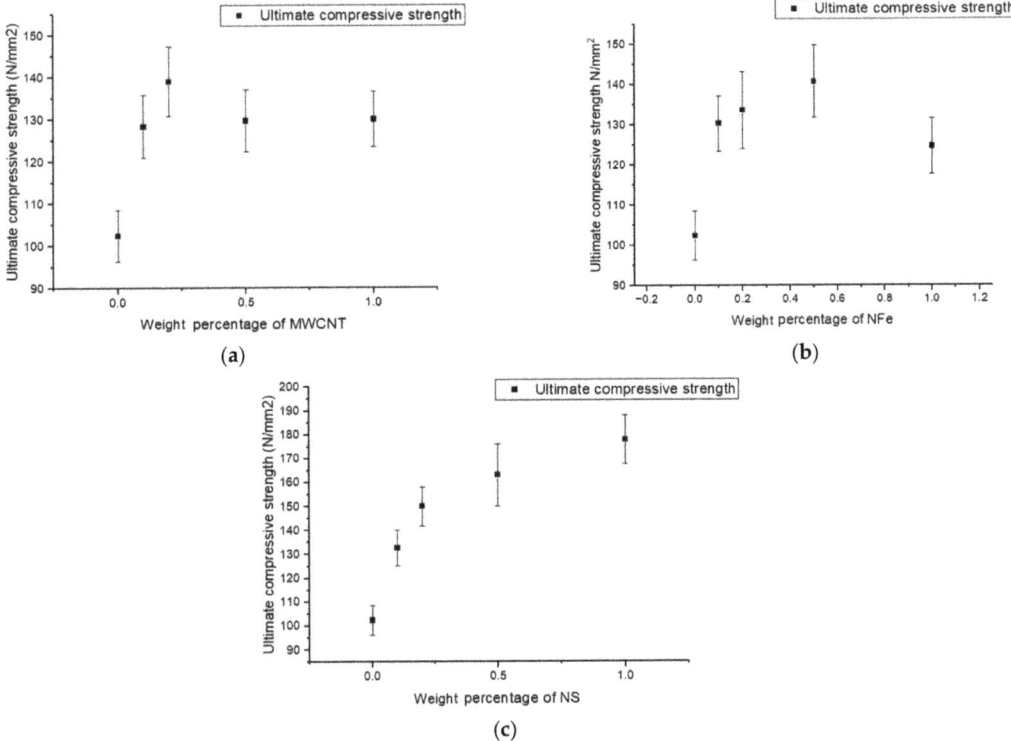

Figure 5. Variation in ultimate compressive strength of modified GFRP samples with different weight percentages of (**a**) MWCNTs, (**b**) NFe, and (**c**) NS.

For the samples with NFe, in Figure 5b, the maximum increase in UCS was observed for the 0.5 wt% sample, which was 37% higher than that of the conventional neat composite, and the minimum was observed in the 1.0 wt% sample, which was still 22% higher than the conventional neat composite. The volume of the nanoparticles is what caused the difference in behavior between the MWCNT sample and the nano-iron oxide samples. Low-volume nano-iron oxide is needed for the same weight of MWCNTs because nano-iron oxide has a higher density. Therefore, compared to the MWCNT samples, a shift from 0.2 wt% to 0.5 wt% was observed for the maximal strength in the nano-iron oxide samples. For the nano-silica set of samples, a parabolic improvement in UCS was observed from the neat conventional composite to the 1.0 wt% sample. The maximum UCS was observed in the maximum wt% sample—that is, the 1.0% sample—and was 74% higher than the conventional composite, as seen in Figure 5c. No decreasing tendency was seen in the other two sets of samples, mainly because of the same molecular formula of glass fiber and nano-silica. This resulted in the addition of supporting reinforcement and glass fibers. The agglomeration that occurred at a high weight percentage was used to distribute the load rather than acting as a stress concentration point.

3.4. Flexural Results

The typical load-deflection curves for the neat GFRP and fiber-reinforced polymer nanocomposite GFRP under 3-point bending are shown in Figure 6. From these graphs, a clear indication of improvement in the flexural modulus of all the fiber-reinforced polymer nanocomposite GFRPs when compared to the neat GFRP composite can be observed from the slope of these curves. The flexural modulus of all the specimens was determined from these graphs using the equation

$$E = \frac{ml^3}{bt^2}$$

where E is the flexural modulus, m is the slope of the curve, b is the width of the specimen, l is the gauge length, and t is the thickness of the specimen. The plastic region for all sets of the fiber-reinforced polymer nanocomposites with MWCNTs is very small, that is, the breaking occurred immediately after the elastic limit as shown in Figure 6a. For the samples of the neat GFRP, there was a prolonged plastic stage before breaking, which is witnessed from the nonlinear regions of the graphs. The maximum improvement in flexural modulus was 87% for the sample with 0.2 wt% of MWCNTs, and the minimum was for the 1.0 wt% MWCNT sample, which was 49.8% higher when compared with the neat GFRP. The energy stored in the sample within the elastic limit is the area under the curve and is 72% higher in the 0.1 wt% MWCNT fiber-reinforced polymer nanocomposite sample when compared with the neat GFRP. In the rest of the samples, even though there was an increase in the flexural modulus, due to early breaking, there was no variation in the energy stored by the samples when compared to the neat GFRP. In the fiber-reinforced polymer nanocomposites with nano-iron oxide, an improvement in the flexural modulus of all the samples similar to the MWCNT FGCMs can be observed in Figure 6b. The flexural modulus improved by 62%, 72%, 66%, and 69% for the 0.1, 0.2, 0.5, and 1.0 wt% samples, respectively, when compared with the neat GFRP sample. Nonlinear regions do exist in these samples, which indicates a bit of plastic deformation before breaking. The energy stored in the nano-iron oxide sample within the elastic limit upon loading was the same as the neat GFRP sample, except for the 1.0 wt% sample, which was 68% higher. Figure 6c shows the load-deflection graphs for the fiber-reinforced polymer nanocomposites with nano-silica and the neat GFRP. Due to the addition of this nano-silica, improvements in flexural modulus, flexural strength, and energy storage were witnessed. The flexural modulus improved by 85% in the 0.5 wt% nano-silica fiber-reinforced polymer nanocomposite sample. The energy stored was 54% higher in the 0.2 wt% nano-silica sample when compared with the neat GFRP sample. The flexural strength of all the samples in the different sets is plotted as shown in Figure 7a–c. From these graphs, it can be said that, with the increase in the weight percentage of MWCNTs, the bending strength increased for the 0.1 wt% and 0.2 wt% samples, and for the remaining two samples, it dropped and remained constant. For the 0.1 wt% and 0.2 wt% MWCNT modified GFRP samples, a 58% and 63% increase in flexural strength, respectively, was observed. Further, for the remaining two sets of samples, the increases in flexural strength were 40% and 43%, respectively. The increase in the flexural properties with the addition of MWCNTs is due to improvements in the compressive properties of the resin. The related fiber-reinforced composite's bending strength rose as a result of this improvement. The change in flexural strength that can be seen in Figure 7b was due to the addition of iron oxide. The percentage increase in flexural strength for these samples, when compared to the unmodified GFRP composite, was 24%, 27%, 47%, and 55%. The following reasons account for the behavior difference between the MWCNT and nano-iron oxide sets of samples: (a) Because the density of iron oxide is higher than that of MWCNTs, the dispersion in epoxy resin is much better in the former than in the latter; and (b) the spherical shape of nano-iron oxide replaces the defects caused by air bubbles, due to which they act as a stress-transfer medium in the laminate. For GFRP modified with NS, the behavior is very similar to the samples modified with nano-iron oxide, which can be observed in Figure 7c. The increase in flexural strength when compared to the conventional composite was 52%, 62%, 64%, and 94% in the order of increasing weight percentage of the samples.

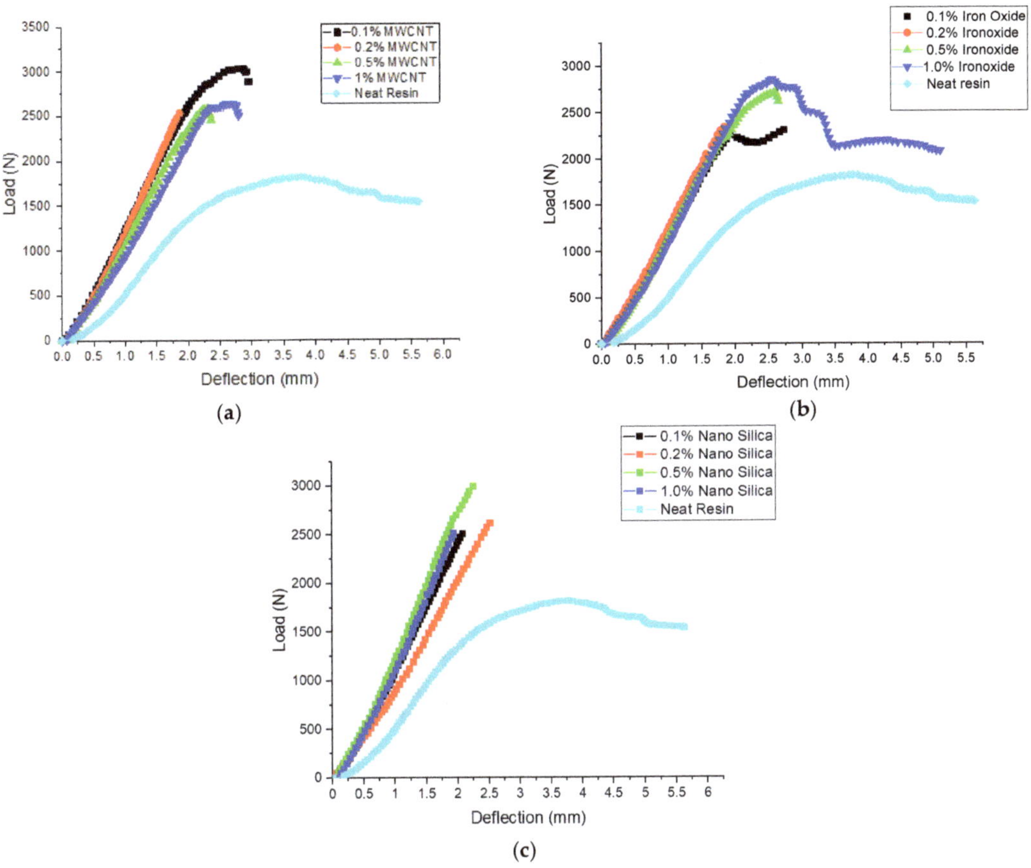

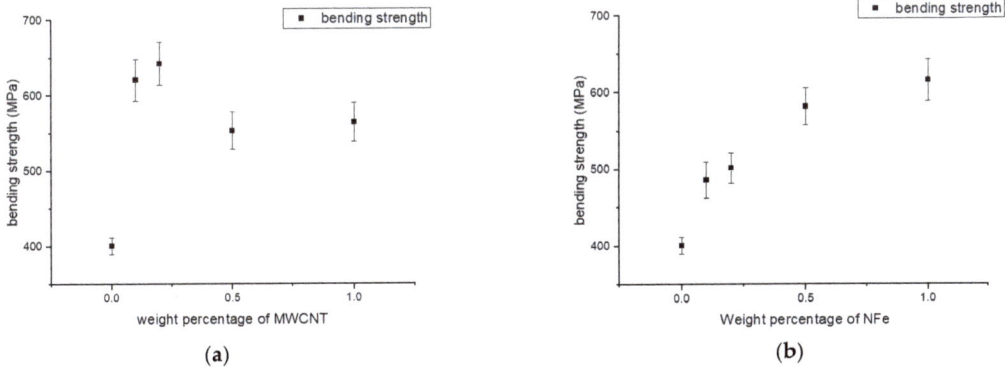

Figure 6. Load vs. deflection graphs of bending tests for modified GFRP with (**a**) MWCNTs, (**b**) NFe, and (**c**) NS.

Figure 7. Cont.

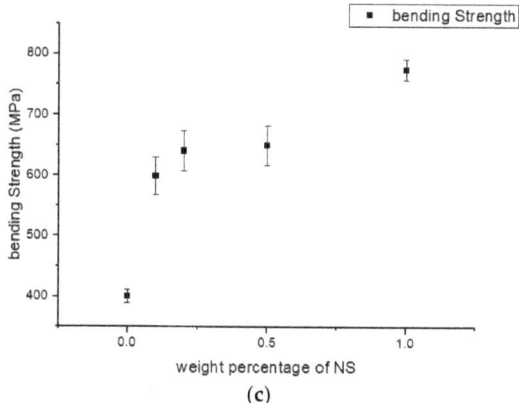

(c)

Figure 7. Variation in flexural (bending) strength of modified GFRP samples with different weight percentages of (**a**) MWCNTs, (**b**) NFe, and (**c**) NS.

3.5. Impact Result

As discussed in Section 3, the Izod impact test was executed, and the impact strength of the samples with diverse nanoparticles such as MWCNTs, NFe, and NS can be seen in Figure 8a–c. For the MWCNT set of samples: (i) the 0.1 wt% sample had the highest impact strength; and (ii) for the 1.0 wt% sample, the rise in impact strength was 66%, and the minimum rise was 26%. The impact strength of the nano-iron oxide sample increased by 127% for the 0.5 wt% sample, reached a maximum of 89% for the 0.1% sample, and reached a minimum of 54% for the 1.0 wt% sample in the case of the samples with nano-silica particles. The escalation in impact strength of the aforesaid lamina is a consequence of morphological modification in the resin during crystallization. The impact strength reduced and remained constant thereafter for the 0.2 wt% sample in the case of the samples with MWCNTs and nano-silica compared to the nano-iron oxide samples because of the agglomeration of these particles in the resin base.

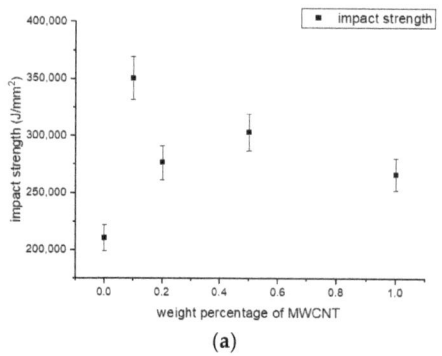

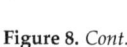

(a)

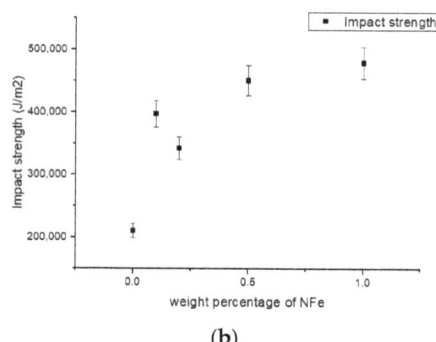

(b)

Figure 8. *Cont.*

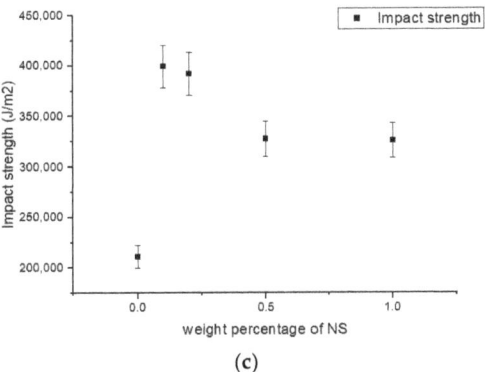

(c)

Figure 8. Impact energy stored under Izod impact test of modified GFRP samples with different weight percentages of (**a**) MWCNTs, (**b**) NFe, and (**c**) NS.

3.6. SEM Analysis

An SEM analysis of the fractured surface of a specimen subjected to tensile and flexural loading was conducted. Figure 9a shows an SEM image of the modified GFRP sample with 0.5 wt% nano-silica, where a fiber pullout phenomenon is not observed and the fibers are broken. This suggests that the presence of nano-silica in the FRP caused an increase in friction for the fibers to come out. Figure 9b shows an agglomeration of nanoparticles, which led to a reduction in tensile strength in the case of the 0.5 wt% MWCNT sample. For the bending-test sample, a rough surface is observed on the fractured area as shown in Figure 9c. This indicates delamination and fiber splitting.

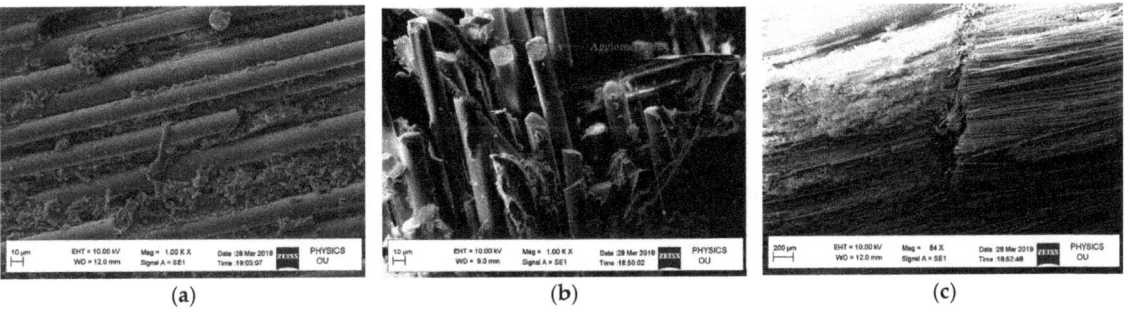

Figure 9. SEM photo of samples: (**a**) tensile load for 0.5 wt% NS, (**b**) tensile load for 0.5 wt% MWCNTs, and (**c**) flexural load for 0.5 wt% NFe.

4. Conclusions

This article studies the mechanical behavior of modified GFRP with nanoparticles with different weight percentages. The maximum tensile strength was found in the 0.5 wt% nano-silica modified GFRP. For all the types of specimens, the ultimate tensile strength decreased with the increasing addition of the nanoparticles because of agglomeration. It is found that the highest percentage gain in compressive strength was 89% for the sample modified with nano-iron oxide (the 0.5 wt% sample). It can also be observed that the nanoscale composites were more brittle than the normal composites, thereby the breaking point was substantially earlier in all samples when compared to the plain resin. With the inclusion of nanoparticles, the impact strength was also enhanced; the largest increase was 127% for the sample modified with nano-iron oxide. In the case of flexural properties, the maximum strength was found in the nano-silica samples with 1.0 wt%. The research on oriented MWCNTs and GF/EP composites reinforced with film-shaped MWCNTs is still ongoing.

Funding: The author extends his appreciation to the Deanship of Scientific Research at King Khalid University for funding this work through the research groups program under grant number R.G.P. 2/129/43.

Institutional Review Board Statement: Not applicable.

Informed Consent Statement: Not applicable.

Data Availability Statement: Not applicable.

Acknowledgments: The author extends his appreciation to the Deanship of Scientific Research at King Khalid University for funding this work through the research groups program under grant number R.G.P. 2/129/43.

Conflicts of Interest: The author declares no conflict of interest.

References

1. Srivasta, V.K.; Pathak, J.P. Friction and wear properties of bushing bearing of graphite filled short glass fibre composites in dry sliding. *Wear* **1996**, *197*, 145–150. [CrossRef]
2. Sutherland, L.S.; Soares, C.G. Impact on low fibre-volume, glass/polyester rectangular plates. *Compos. Struct.* **2005**, *68*, 13–22. [CrossRef]
3. Evci, C.; Gulgec, M. An experimental investigation on the impact response of composite materials. *Int. J. Impact Eng.* **2012**, *43*, 40–51. [CrossRef]
4. Sutherland, L.S.; Soares, C.G.; Guecdes, C. Mechanical properties of MWCNTs and graphene nanoparticles modified glass fibre-reinforced polymer nanocomposite. *Int. J. Impact Eng.* **2005**, *33*, 194.
5. Mitreveski, T.; Marshall, I.H.; Thomsan, R. The influence of impactor shape on the damage to composite laminates. *Compos. Struct.* **2006**, *76*, 116–122. [CrossRef]
6. Chang, M.S. An investigation on the dynamic behavior and thermal properties of MWCNTs/FRP laminate composites. *J. Reinf. Plast. Compos.* **2010**, *29*, 3593–3599. [CrossRef]
7. Lal, A.; Markad, K. Thermo-mechanical post buckling analysis of multiwall carbon nanotube-reinforced composite laminated beam under elastic foundation. *Curved Layer. Struct.* **2019**, *6*, 212–228. [CrossRef]
8. Ayatollahi, M.R.; Shadlou, S.; Shokrieh, M.M.; Chitsazzadeh, M. Effect of multi-walled carbon nanotube aspect ratio on mechanical and electrical properties of epoxy-based nanocomposites. *Polym. Test.* **2011**, *30*, 548–556. [CrossRef]
9. Rafiee, M.; Nitzsche, F.; Labrosse, M.R. Fabrication and experimental evaluation of vibration and damping in multiscale graphene/fiberglass/epoxy composites. *J. Compos. Mater.* **2019**, *53*, 2105–2118. [CrossRef]
10. Rafiee, M.; Nitzsche, F.; Laliberte, J.; Thibault, J.; Labrosse, M. Simultaneous einforcement of matrix and fibers for enhancement of mechanical properties of graphene-modified laminated composites. *Polym. Compos.* **2019**, *40*, E1732–E1745. [CrossRef]
11. Mostovoy, A. Fiber-Reinforced Polymer Composites: Manufacturing and Performance. *Polymers* **2022**, *14*, 338. [CrossRef] [PubMed]
12. Vázquez-Moreno, J.M.; Sánchez-Hidalgo, R.; Sanz-Horcajo, E.; Viña, J.; Verdejo, R.; López-Manchado, M.A. Preparation and Mechanical Properties of Graphene/Carbon Fiber-Reinforced Hierarchical Polymer Composites. *J. Compos. Sci.* **2019**, *3*, 30. [CrossRef]
13. Bekeshev, A.; Mostovoy, A.; Tastanova, L.; Kadykova, Y.; Kalganova, S.; Lopukhova, M. Reinforcement of Epoxy Composites with Application of Finely-ground Ochre and Electrophysical Method of the Composition Modification. *Polymers* **2020**, *12*, 1437. [CrossRef] [PubMed]
14. Jamali, N.; Rezvani, A.; Khosravi, H.; Tohidlou, E. On the mechanical behavior of basalt fiber/epoxy composites filled with silanized graphene oxide nanoplatelets. *Polym. Compos.* **2018**, *39*, E2472–E2482. [CrossRef]
15. Tian, Y.; Zhang, H.; Zhang, Z. Influence of nanoparticles on the interfacial properties of fiber-reinforced-epoxy composites. *Compos. Part A Appl. Sci. Manuf.* **2017**, *98*, 1–8. [CrossRef]
16. Liang, S.; Jia, C.; Tang, Y.; Zhang, Y.; Zhang, J.; An-qiang, P. Mechanical and electrical properties of carbon nanotube reinforced epoxide resin composites. *Trans. Nonferrous Met. Soc. China* **2007**, *17*, 675–679.
17. Zhou, Y.; Pervin, F.; Lewis, L.; Jeelani, S. Fabrication and characterization of carbon/epoxy composites mixed with multi walled carbon nanotubes. *Mater. Sci. Eng. A* **2008**, *475*, 157–165. [CrossRef]
18. Wong, M.; Paramsothy, M.; Xu, X.J.; Ren, Y.; Li, S.; Liao, K. Physical interactions at carbon nanotube–polymer interface. *Polymer* **2003**, *44*, 7757–7764. [CrossRef]
19. Veronica, M.M.; Veronica, P.S.; Jose, M.M.M. Improvement in mechanical and structural integrity of natural stone by applying unsaturated polyester resin-nanosilica hybrid thin coating. *Eur. Polym. J.* **2008**, *44*, 3146–3155.

20. Shu, X.Z.; Li, M.W.; Jian, S. Effect of Nanosilica on the properties of Polyester-Based Polyurethane. *J. Appl. Polym. Sci.* **2003**, *88*, 189–193.
21. Smia, G.; Shailesh, K.G.; Lon, J.M. Surface modification of nano-silica with amides and imides for use in polyester nano composites. *J. Mater. Chem. Am.* **2013**, *1*, 6073–6080.
22. Natalia, C.C.; Jose Roberto, V.B.; Yendry, C.U. Basis and Applications of Silicon Reinforced Adhesives. *Org. Med. Chem Int. J.* **2018**, *5*, 18–29.
23. Sudirman, M.; Anggaravidya, E.; Budianto, I. Gunawan Synthesis and Characterization of Polyester-Based Nanocomposite. *Procedia Chem.* **2012**, *4*, 107–113. [CrossRef]
24. Zheng, Y.; Ning, R.; Zheng, Y. Study of SiO2 nanoparticies on the improved performance of epoxy and fibre composites. *J. Reinf. Plast. Compos.* **2005**, *24*, 223–233. [CrossRef]

Article

Flame Retardant Behaviour and Physical-Mechanical Properties of Polymer Synergistic Systems in Rigid Polyurethane Foams

Branka Mušič *[ ], Nataša Knez and Janez Bernard

Slovenian National Building and Civil Engineering Institute, 1000 Ljubljana, Slovenia
* Correspondence: branka.music@zag.si; Tel.: +386-1-2804-370

Abstract: In the presented work, the influence of two flame retardants—ammonium polyphosphates and 2,4,6-triamino-1,3,5-triazine on the polyurethane foam (PUR) systems were studied. In this paper, these interactive properties are studied by using the thermal analytical techniques, TGA and DTA, which enable the various thermal transitions and associated volatilization to be studied and enable the connection of the results with thermal and mechanical analysis, as are thermal conductivities, compression and bending behavior, hardness, flammability, and surface morphology. In this way, a greater understanding of what the addition of fire retardants to polyurethane foams means for system flammability itself and, on the other hand, how this addition affects the mechanical properties of PUR may be investigated. It was obtained that retardants significantly increase the fire resistance of the PURs systems while they do not affect the thermal conductivity and only slightly decrease the mechanical properties of the systems. Therefore, the presented systems seem to be applicable as thermal insulation where low heat conductivity coupled with high flame resistance is required.

Keywords: flammability; polyurethane polymer; foams; thermal conductivity; mechanical properties

Citation: Mušič, B.; Knez, N.; Bernard, J. Flame Retardant Behaviour and Physical-Mechanical Properties of Polymer Synergistic Systems in Rigid Polyurethane Foams. *Polymers* **2022**, *14*, 4616. https://doi.org/10.3390/polym14214616

Academic Editor: S. D. Jacob Muthu

Received: 16 September 2022
Accepted: 25 October 2022
Published: 31 October 2022

Publisher's Note: MDPI stays neutral with regard to jurisdictional claims in published maps and institutional affiliations.

Copyright: © 2022 by the authors. Licensee MDPI, Basel, Switzerland. This article is an open access article distributed under the terms and conditions of the Creative Commons Attribution (CC BY) license (https://creativecommons.org/licenses/by/4.0/).

1. Introduction

Polyurethane foams are materials known for a long time, the use of which continues to increase over the years. In addition to the already known wide applicability in various industries such as construction, e.g., for thermal insulation in windows and doors and fastening and sealing of joinery, in the automotive industry, as well as in households for various purposes, for fastening fence posts in the garden and as an electrical insulator, etc., recently it is even used in geotechnical applications for soil reinforcement [1] and as an insulating material for building walls, e.g., in attics. With increasing use, there is also increasing interest in potential improvements of this material and, consequently, also interest in the influence of various additives, especially in terms of reaction to fire and on the physical-mechanical properties of polyurethanes. Polyurethane foams are materials with a low weight-to-strength ratio, low electrical conductivity as well as low heat conductivity [2]. On the other hand, the downside of PURs is their high flammability [3].

Due to increased climate and environmental concerns, there was a need to design a new effective fire-retardant system from halogen-free fire retardants [4–7], with aluminum hydroxide (ATH), magnesium hydroxide (MDH), carbon nanotube (CNT), expandable graphite, halloysite nanotubes with POSS, etc. Furthermore, in the construction industry, a lot of effort has been put to fulfil ever stricter reactions to fire requirements, and therefore over the years a lot of knowledge has been accumulated about the influence of various environmentally acceptable additives and retardants to fire resistance of the PURs, with a synergistic effect on improving the thermal insulating properties [8,9]. However, there is less known about the influences of those fire-retardant additives on the mechanical properties of the PURs.

In the presented study, the influence of two compounds: the ammonium polyphosphates and 2,4,6-Triamino-1,3,5-triazine on fire resistance, thermal conductivity, and different mechanical properties of PURs systems were systematically investigated.

2. Materials and Methods

2.1. Materials

- A two-component polyurethane foam "Tekapur Polefix" (PURs), TKK d.o.o. (Srpenica, Slovenia) Component A is a polyol with several hydroxyl groups and triethyl phosphate. Component B is a polymethylene polyphenyl polyisocyanate.
- Ammonium polyphosphate, Exolite AP 422 (APP) was supplied by Clariant (Mutenz, Switzerland), it is Ammonium polyphosphate, white fine powder, non-hygroscopic, non-flammable, halogen-free, with bulk density 700 kg/m^3, and melting point ~240 °C (decomposition).
- 2,4,6-Triamino-1,3,5-triazine (TATA), Sigma Aldrich (St. Louis, MO, USA), white powder, with bulk density 800 kg/m^3, and melting point ~354 °C (decomposition).

2.2. Preparation of PURs

All PURs were individually prepared according to the same procedure and using a mold in which PUR foam expanded. First, the appropriate amount of PUR component B and a flame retardant (except for PUR 0) were weighted into a mixing vessel and mixed with high-speed mechanical stirrer, at about 1400 rpm for 10 min to obtain a homogeneous mixture. After that, the appropriate amount of component A was poured into the mixture which was further homogenized with a stirrer at 1000 rpm and transferred into a mold with enough free space to enable the full expansion of the foam during curing. After about 45 s, the foam begins to expand. The foam reaches the final volume in about three minutes and after 10–15 min. The foams were allowed to cure for 72 h, at room conditions T = 23 ± 2 °C and relative humidity 50 ± 15 % in accordance with ISO 291:2008. After curing, the foams were cut into standard shaped specimens for further testing. When preparing the samples, we made sure that the samples were as uniform as possible.

The structure and resulting performance of polyurethane foams are driven by the stoichiometry of the polymerization reaction, which is directly impacted by applied monomers, additives, their chemical composition, and the ratio between the polyols and isocyanates [10]. The amount of hydroxyl and isocyanate groups present in the system are essential for reactions leading to the generation of urethane bonds [11,12].

The reference PUR without additions was designated as PUR 0; the foam with addition of APP was designated as PUR 1 and finally, the foam with the addition of TATA was designated as PUR 2. The ratio used in PUR 0 between polyol and isocyanate was according to the manufacturer's recommendations, therefore the mixing weight ratio was 1:1.22. From preliminary research, we found that a maximum of 30% of the fire-retardant additive can be included in the system, based on the total weight of the A + B component, so that the fire retardant powder is homogeneously mixed into the B component, the expansion takes place on the scale of PUR 0 and the polymerization reaction ends (mass is not sticky after expansion). In Table 1 the contents of raw materials in PURs are given.

Table 1. Compositions of the PURs.

Specimens	Raw Material Ratio (%)	Component A (g)	Component B (g)	APP (g)	TATA (g)
PUR 0	Comp. A:Comp. B = 45:55	32.72	40	/	/
PUR 1	Comp. A:Comp. B:APP = 34.61:42.31:23.08	32.72	40	21.82	/
PUR 2	Comp. A:Comp. B:TATA = 34.61:42.31:23.08	32.72	40	/	21.82

2.3. Methods of Characterization

Unless stated otherwise, before characterization the specimens were conditioned for at least 24 h at standard laboratory conditions at 23 ± 2 °C and 50 ± 5% relative humidity. Further on, mechanical properties and the apparent densities were determined at stated conditions also, as required by relevant standards. The published mechanical properties and apparent densities are presented as the average of the 5 measurements ± standard deviations, while other characteristics were obtained on single specimen measurement.

2.3.1. Apparent Densities

The apparent densities of the PUR specimens were determined according to ISO 845:2006. The dimensions of the specimens were (50 mm × 50 mm × 50 mm) ± 1 mm.

2.3.2. Thermal Conductivity

The thermal conductivity of the PUR specimens was determined in a home-made heat flow setup. Prior to testing the specimens were conditioned at 70 °C for 14 days and further two days at 23 °C, 50% RH. The dimensions of the specimens were (100 mm × 60 mm × 10 mm) ± 1 mm. Thermal conductivity was determined on the specimens inserted in-between cold and hot plates with temperatures of 15 °C and 25 °C, respectively.

2.3.3. Thermal Decomposition

Thermal decomposition of the PUR specimens was determined with thermogravimetric analysis (TG), Netzsch instrument STA 409PC Luxx, Weyhe, Germany. The specimens with a mass of about 25 mg were heated in airflow from room temperature to 900 °C with rate of 10 K/min.

2.3.4. Compression and Bending Behavior

The compression and bending behavior of the PUR specimens were determined on a universal test machine Zwick Z030, Zwick Roell Group, Ulm, Germany. Compression properties were determined according to EN 826:2013. The test specimens of dimensions (50 mm × 50 mm × 50 mm) ± 1 mm were compressed between the two plates of the universal test machine and at a constant rate of 0.5 mm/min was applied to the specimen till failure occurred. Bending behavior was determined according to the requirements of EN12089:2013. The specimens of dimensions of (150 mm × 30 mm × 50 mm) ± 1 mm were tested in three-point bending mode in a universal test machine. At a constant rate of 0.5 mm/min till failure occurred.

2.3.5. Hardness

The hardness of the samples was measured using a device known as a Durometer and the determined hardness values are therefore referred to as durometer hardness. Durometer hardness is a dimensionless quantity; it represents a relative comparison of hardness between different, yet similar grades of materials, having hardness measured on the same durometer scale. The Shore A hardness tester (Zwick, Ulm, Germany) was used for determining the hardness of PUR samples, according to EN ISO 868:2004. For each sample, eight measurements were taken.

2.3.6. Flammability

Flammability of the PURs were obtained according to UL-94 HB on the specimens with dimensions of 125 mm × 15 mm × 100 mm) ± 1 mm. A Horizontal burning test was performed.

2.3.7. Cone Calorimetry

Reactions to fire properties were studied by using a cone calorimeter, produced by Fire Testing Technology, East Grinstead, UK according to ISO 5660-1:2015. Specimens were exposed to a heat flux of the 40 kW/m^2.

2.3.8. Loss of Ignition Test (LOI)

Loss on ignition was assessed from the weight of the test specimens before and after the exposure to the 40 kW/m^2 in a cone calorimeter.

2.3.9. FTIR Analysis

Exhaust gases released during exposure of the test specimens to the 40 kW/m^2 in the cone calorimeter were analyzed by means of FTIR analyzer atmosFIR produced by Protea, Middlewich, UK according to ISO 19702:2015.

2.3.10. Surface Morphology

The distribution of solid flame-retardant particles and the shape and size of sample porosity was observed using a scanning electron microscope (SEM) JSM-IT500LV, Oxford Inca; Jeol, Oxford Instruments Analytical (Freising, Germany), with an integrated energy-dispersive spectroscopy, W filament, fully automatic gun alignment, and in low (10–650 Pa) vacuum mode.

3. Results

3.1. Apparent Densities

The densities of the PURs are as follows PUR 0 (46.70 kg/m^3), PUR 1 (62.61 kg/m^3), and PUR 2 (60.04 kg/m^3). The densities of PUR 1 and PUR 2 are comparable and about 30% higher than PUR 0. For PUR 1 and PUR 2, the same amount of additive 25 mass % was added.

3.2. Thermal Conductivity

The thermal conductivity of the specimens are as follows PUR 0 (36.5 mW/mK), PUR 1 (36.4 mW/mK), PUR 2 (35.6 mW/mK). Presented values correspond well to apparent densities of the specimens as a higher density of the cellular insulation generally contributes to increasing its thermal conductivity.

3.3. Thermal Decomposition

Thermal decomposition curves (TG) are presented in Figure 1. The mass losses at the first decomposition step were 36.3 wt.% for PUR 0, 28.3 wt.% for PUR 1, and 51.5 wt.% for PUR 2. Respectively, the mass losses for the second step were 55.6 wt.%, 65.4 wt.%, and 46.0 wt.%. Decomposition steps end at 338 °C and 662 °C for PUR 0, at 293 °C and 815 °C for PUR 1 and at 336 °C and 638 °C for PUR 2. TG curves of PUR specimens are presented in a Figure 1.

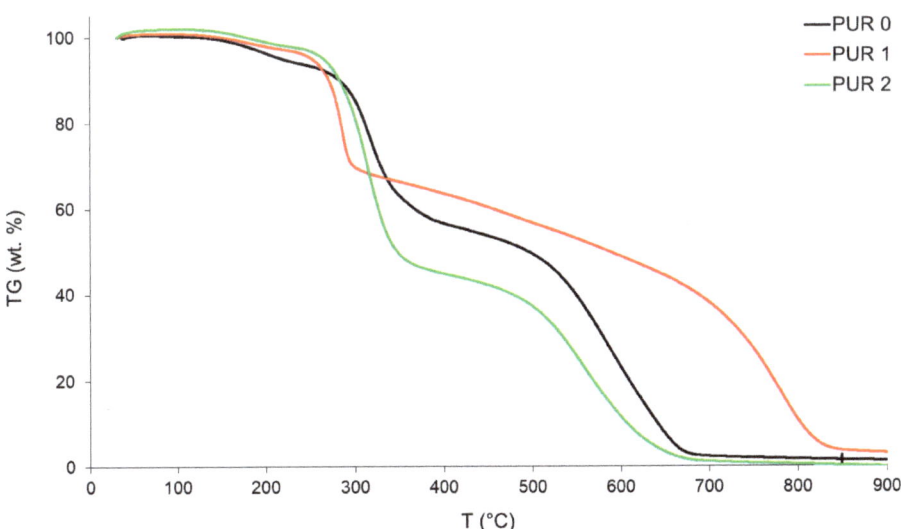

Figure 1. TG curves of the PURs.

The first decomposition steps of all the PUR specimens were completed in a relatively narrow temperature range in-between 293 °C and 338 °C.

From Figure 1 we can see also that the course of the weight loss curve is similar for PUR 0 and PUR 2, while different for PUR 1. The weight loss is slightly lower at PUR 1 and PUR 2 than at PUR 0 up to a temperature of 293 °C. In PUR 1 was added APP, a high molecular weight phosphate-based chain, it serves as both an acid source and a blowing agent in intumescent formulations known to promote char formation during polymer decomposition. At elevated temperatures, the phosphorus containing the flame-retardant additive, APP, decomposes to produce phosphoric and polyphosphoric acids, which consequently promote charring via cross-linking of reactive polymer fragments [13]. The formation of carbonized char networks prevents or slows the transfer of heat, oxygen, and combustible volatiles into the pyrolysis zone; hence retarding the flaming/combustion process. Detailed mechanistic schemes describing the charring behavior of APP containing resin formulations have been discussed by Kandola and Ullah [13,14]. Values of partial weight loss in sample PUR 2 in lower temperature ranges are related to water evaporation. Weight loss at around 336 °C in PUR 2 was also due to partial loss of formaldehyde, methanol, and amine. The polycondensation reaction of melamine took place at temperatures above 336 °C when the products underwent a number of independent reactions involving both side chain and ring degradation. This means that some melamine molecules can be sublimated at a temperature lower than the sublimation temperature typically observed at 345 °C. Weight loss also occurs due to the release of formaldehyde, methanol, amine, and NH_3 from melamine (at about 390 °C). Weight loss at temperatures above 450 °C involves the general thermal decomposition of melamine, which ends above 660 °C with the decomposition of melamine to form volatile products, including CO_2, HCN, and CO [15].

3.4. Compression and Bending Behaviour

The compressive properties were determined on three parallel samples. Figure 2 show how the deformation of PUR 0, PUR 1 and PUR2 varied continuously with increasing standard force. The PUR 0 and PUR 2 samples behave similarly, while the PUR 1 sample has slightly worse result.

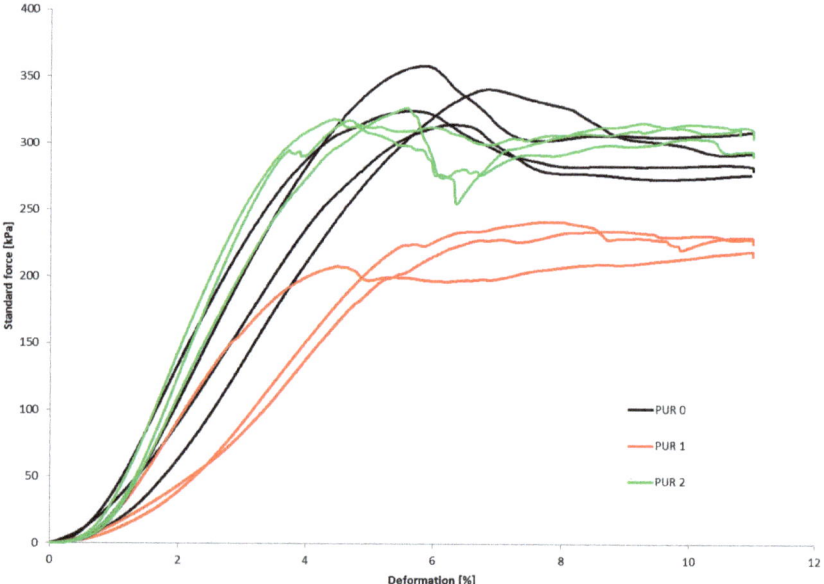

Figure 2. Compressive behavior of samples PUR 0, PUR 1 and PUR 2.

The bending properties were determined on three parallel samples. Figure 3 show how the deformation of PUR 0, PUR 1 and PUR2 varied continuously with increasing standard force.

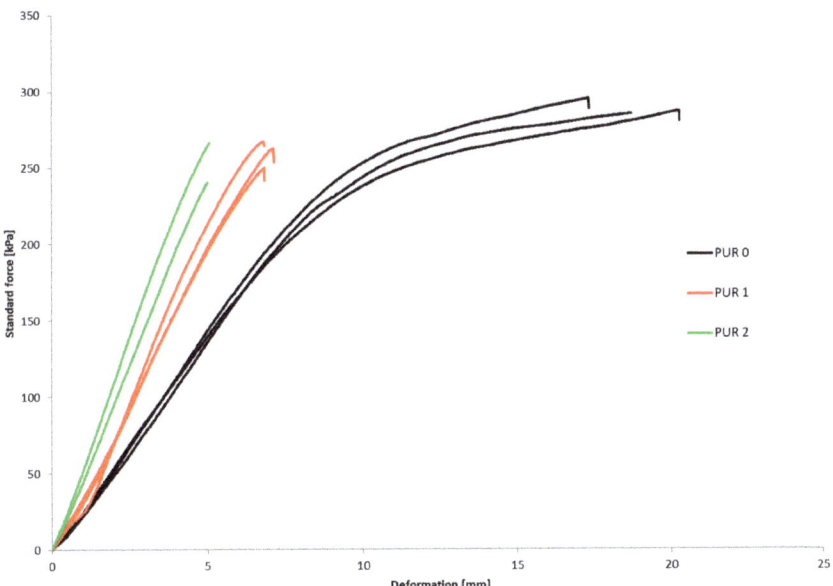

Figure 3. Bending behavior of samples PUR 0, PUR 1 and PUR 2.

The compressive and bending properties of the specimens are summarized and presented in Table 2. The (σ_M) represents compressive strength and (σ_b) bending strength.

Table 2. Mechanical properties of PURs.

Specimens	σ_M (MPa)	σ_b (MPa)
PUR 0	335 ± 19	293 ± 6
PUR 1	220 ± 11	260 ± 9
PUR 2	314 ± 16	253 ± 19

3.5. Hardness

The Shore A scale is employed for softer/flexible materials. The measured values indicate the resistance to indentation of the tested material on a scale between 0 and 100.

The hardness test is based on the measurement of the penetration of a rigid peak into the specimen under specified conditions. The measured penetration is converted into International Rubber Hardness Degrees (IRHD). The hardness scale of degrees is chosen such that 0 represents a material having an elastic modulus of zero, and 100 represents a material of infinite elastic modulus.

Table 3 shows the durometer hardness of PURs samples, whereas a thumb rule, higher numbers on the scale indicate a greater resistance to indentation, which means harder material.

Table 3. Hardness of PURs.

Specimens	Durometer Hardness (Shore A)
PUR 0	31 ± 4
PUR 1	16 ± 4
PUR 2	14 ± 3

3.6. Flammability

Samples PUR 0, PUR 1, and PUR 2 were prepared and tested for combustion in accordance with the UL-94 HB standard. The results are shown in Table 4. The dripping of samples during burning did not occur in any case. The samples stopped burning immediately after removing the fire source. The fire reached the marked line of the PUR 0 after burning 30 s, whereas PUR 1 and PUR 2 preserved more unburning material than neat PUR 0. It can be observed that in the case PUR 1 and PUR 2 promoted the formation of a compact burned layer.

Table 4. Burnt PURs samples in accordance with UL94 standard.

Specimens	V (mm/min)	L (mm)	t (s)	Flame Passed 25 mm Mark	Flame Passed 100 mm Mark	Flammability
PUR 0	35	35	60	Yes	No	Slowly self-extinguishing after withdrawal of fire
PUR 1	20	20	60	No	No	Fast self-extinguishing after withdrawal of fire
PUR 2	15	15	60	No	No	Fast self-extinguishing after withdrawal of fire

V is linear burning rate in mm/minute (mm/min); L is the damaged length, in millimeters (mm); t is time, in seconds (s).

Visual differences between PURs, according to the UL-94 HB burning test, are presented for one set in Figure 4. We can see that in PUR 0 the line is no longer visible, in the case of PUR 1 with the fire-retardant additive APP the burning has reached the line, and in the case of PUR 2 the flame retardant TATA works even better.

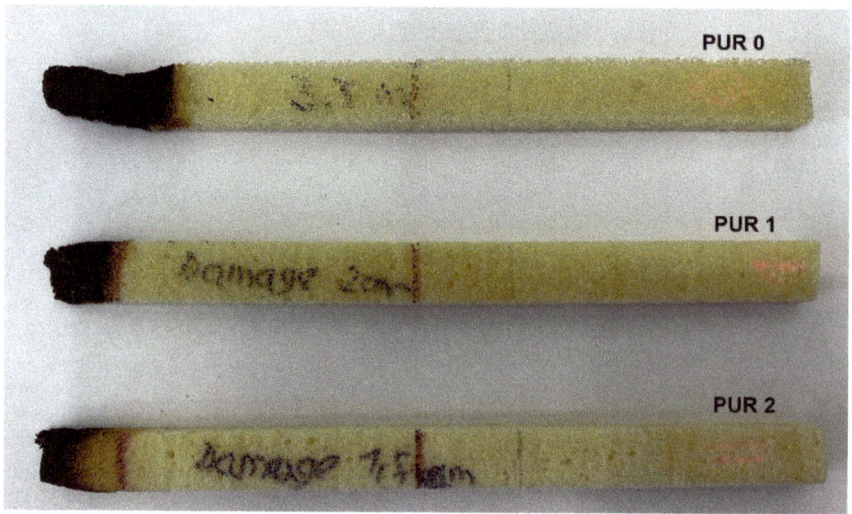

Figure 4. Visual differences between PUR 0, PUR 1, and PUR 2 after the UL-94 HB burning test.

Cellular material PUR 0 burns readily in the presence of oxygen and heat with a very high fire spread rate and a high smoke release rate. As per the experimental evidence, the pores of the foam entrap air further aid in its combustion [16,17].

The mechanism of APP (added to PUR 1) degradation has been investigated and consists of the release of water and ammonia and the formation of polyphosphoric acid, which is then volatilized and dehydrated at temperatures above 250 °C [18]. APP is also thought to promote an intumescent layer of char, which acts as a physical barrier to slow the mass transfer of heat. Due to both processes, the halogen-free flame-retardant APP is considered a very effective phosphorus-based flame retardant used in polymers because it is more environmentally friendly, highly effective, and low in toxicity. However, it is necessary to note, as can be seen from the results, that APP can affect the deterioration of the physical and mechanical properties of the composite and increase the generation of smoke [19–22].

In the PUR 2 sample, the flame-retardant melamine captures the heat of the PUR matrix during combustion and undergoes advanced endothermic condensation with the evolution of ammonia. In the first phase, water from the sample evaporates. This is followed by the breaking of urethane bonds and decomposition. Melamine does not begin to decompose until somewhere above 450 °C and involves general thermal decomposition of melamine, which ends above 660 °C [16]. According to the UL-94 HB test, we can see that PUR 2 had the best results.

3.7. Cone Calorimetry

Cubes of 50 mm made of the three PURs were cut in 10 mm thick squares. From each type of PUR two 100 mm × 100 mm, 10 mm thick specimens were prepared, two of each type of PUR. Specimens were exposed to the 40 kW/m^2 heat flux in a cone calorimeter. For each type of PUR, a self-ignition of the exposed specimen was observed as well as ignition of the exposed specimen initiated by sparks. The heat release rate and total heat release parameters were compared for the three PURs under both conditions—without or with a help of a spark igniter [23–26]. Table 5 shows the appearance of PUR 0, PUR 1 and PUR 2 samples before, during and after the test.

From Figures 5 and 6 we can see that all specimens ignited in a few seconds after the heat flux exposure. For PUR 1 and PUR 2 specimens a white smoke has been noticed before ignition. PUR 1 and PUR 2 specimens expanded as seen in the photo. After the test of PUR 0, only a very small number of residuals were left whereas PUR 1 and PUR 2 specimens still had a firm structure.

During heat flux exposure in the cone calorimeter, all specimens ignited in both ignition modes. From Table 6 we can see that the specimens ignited faster when a spark igniter was used.

The greatest difference in ignition time was for PUR 0. PUR 1 self-ignited fast, but the heat release rate was significantly lower compared to HRR when the ignition was induced with the spark igniter. The heat release rate for PUR 2 specimen was similar for both modes of ignition, for self-ignition and ignition with a spark igniter. In both PUR 2 specimens, the HRR curve has two peaks. In addition, THR is similar for the two ignition modes for PUR 2 whereas for PUR 0 and PUR 1 the THR is significantly lower when specimens were self-ignited.

The smoke production rate (Figures 7 and 8) was similar for both ignition modes in all three PUR types. Smoke production was the greatest for PUR 0 specimens and the lowest for PUR 2 specimens [27]. Two peaks were noticed in PUR 2 specimens.

Table 5. Prepared test specimens of the PURs, before, during and after exposure to the 40 kW/m² of heat flux in cone calorimeter.

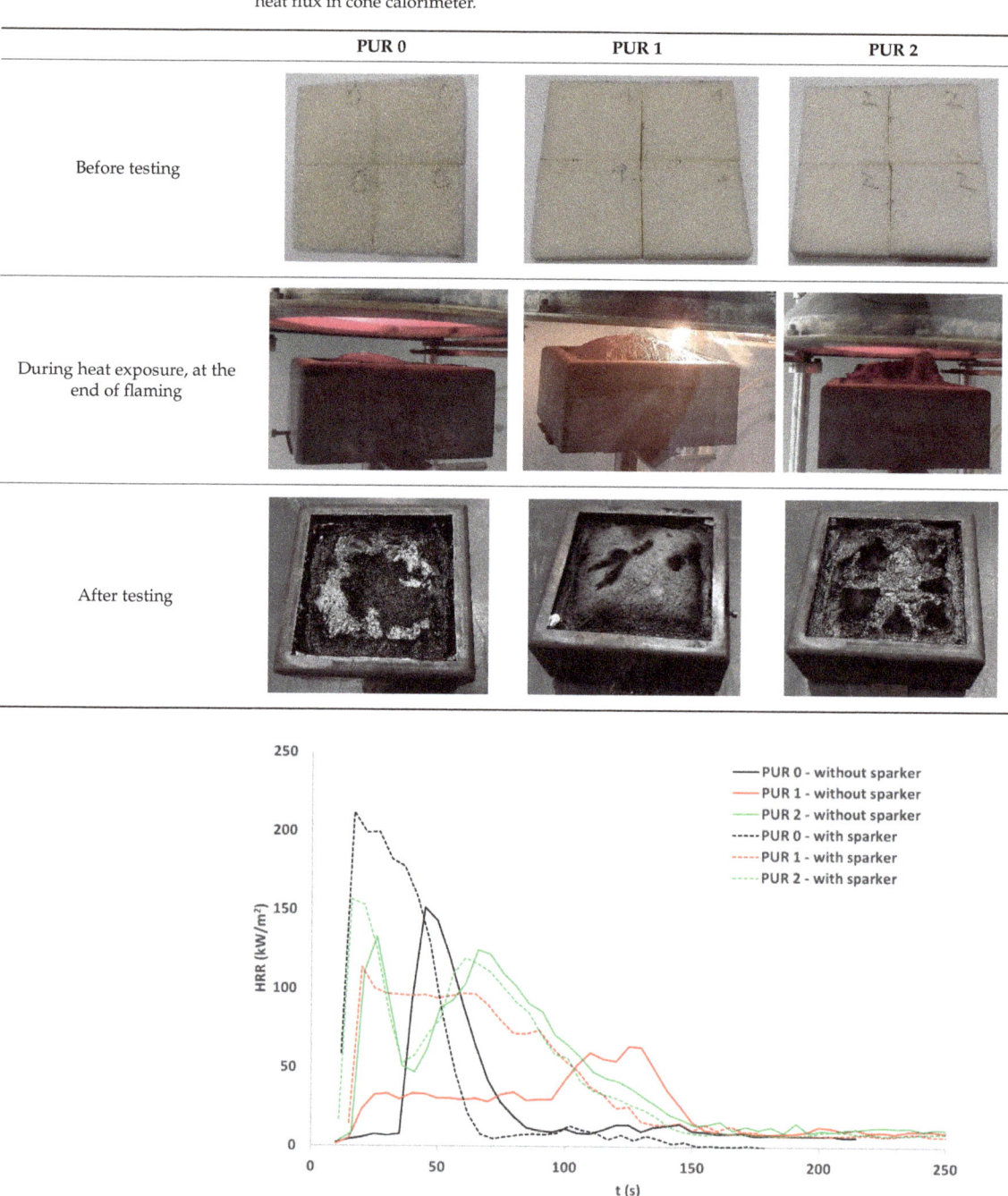

Figure 5. Heat release rate (HRR) of the PURs samples exposed to the 40 kW/m² of heat flux; self-ignition and spark ignition.

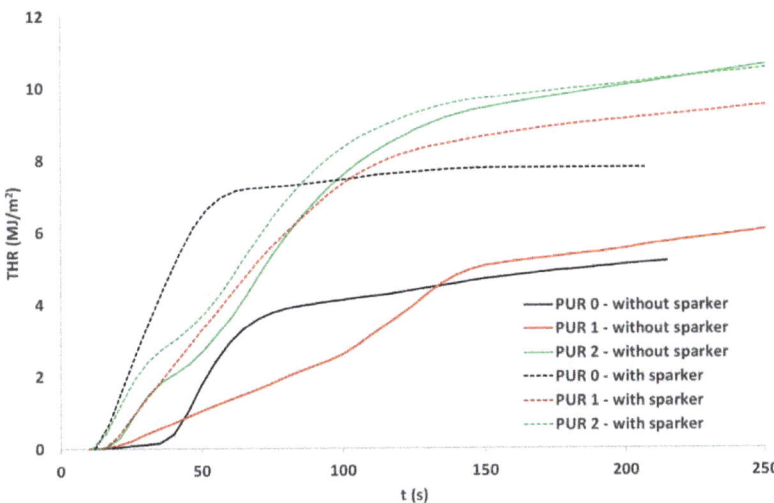

Figure 6. Total heat release (THR) of the PURs exposed to the 40 kW/m^2 of heat flux; self-ignition and spark ignition.

Table 6. Ignition times and flame duration for the three PURs exposed to 40 kW/m^2 of heat flux self-ignited or ignited with a spark igniter.

Specimens	Ignition	Ignition Time (s)	Flame Duration (s)
PUR 0	Self-ignition	33	48
	Ignition with sparks	3	57
PUR 1	Self-ignition	10	150
	Ignition with sparks	3	165
PUR 2	Self-ignition	12	130
	Ignition with sparks	6	134

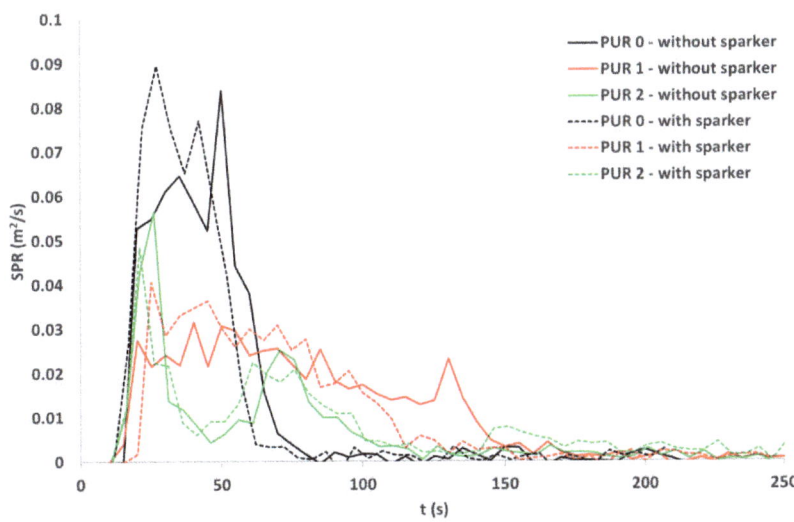

Figure 7. Smoke production rate (SPR) of the PURs samples exposed to the 40 kW/m^2 of heat flux; self-ignition and spark ignition.

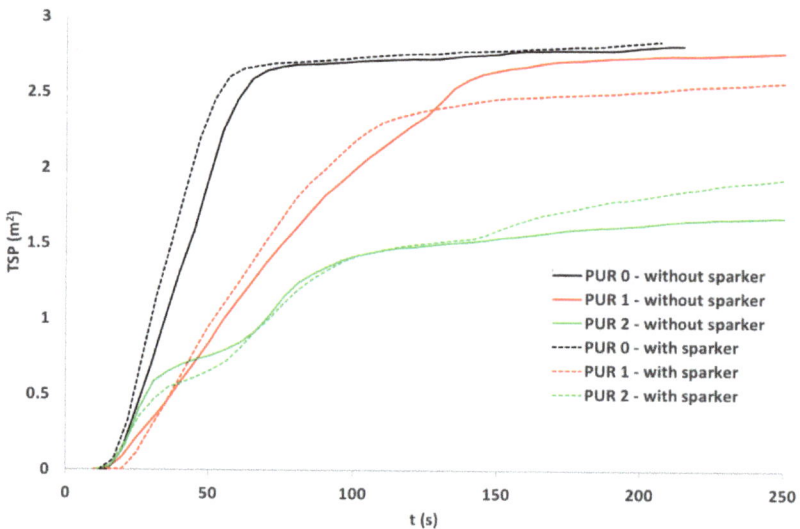

Figure 8. Total smoke production (TSP) of the PURs samples exposed to the 40 kW/m² of heat flux; self-ignition and spark ignition.

3.8. Loss of Ignition Test (LOI)

During exposure of the three PUR specimens to the 40 kW/m² in a cone calorimeter, mass loss of the specimens was measured. We can see the mass loss shown in Figure 9. Two modes of ignition were compared for the three PURs—self-ignition and ignition with sparks. The difference between the mass before the test and the final mass after heat flux exposure was calculated for each PUR type and ignition mode. Loss on ignition was calculated as a percentage of mass loss compared to initial mass.

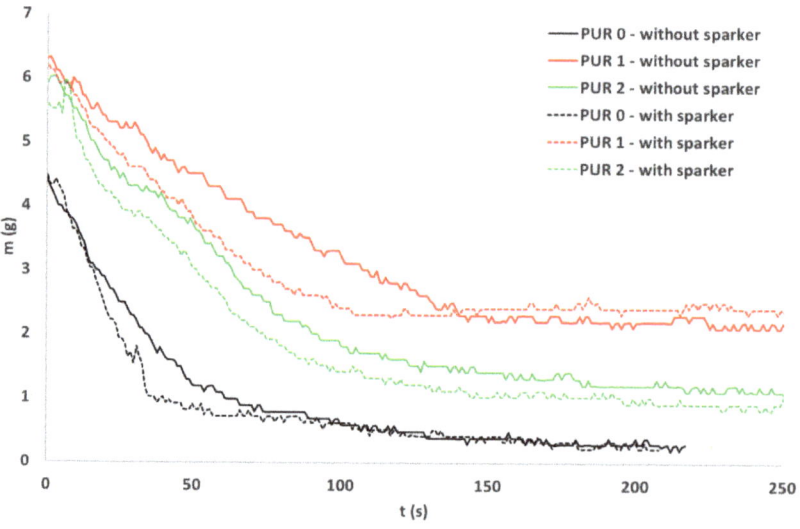

Figure 9. Mass loss of PUR samples exposed to the 40 kW/m² heat flow; self-ignition and spark ignition.

Loss on ignition was similar for the two ignition modes for all three PURs. On the other hand, LOI differs between different PURs, in PUR 0 LOI was around 95%, in PUR 1 around 65%, and in PUR 2 around 80%, as shown in Table 7.

Table 7. Initial mass and loss on ignition for PURs exposed to the 40 kW/m² of heat flux; self-ignited and ignited with a spark igniter.

Specimens	Ignition	Initial Mass (g)	LOI (%)
PUR 0	Self-ignition	4.4	94
	Ignition with sparks	4.5	95
PUR 1	Self-ignition	6.3	67
	Ignition with sparks	6.1	61
PUR 2	Self-ignition	5.92	81
	Ignition with sparks	5.61	83

3.9. FTIR Analysis

During heat flux exposure of test specimens in a cone calorimeter, the exhaust gases were continuously analyzed by an FTIR analyzer. Concentrations of several gases were calculated from IR spectra [28]. Such calculations can lead to certain inaccuracy, especially where the amount of certain gas is low or other gases with a similar spectrum are present. Negative values on graphs are the result of characteristics of the calculation method.

Concentrations of several gases were calculated. Concentrations of carbon dioxide (Figure 10) and carbon monoxide (Figure 11) were measured with a cone calorimeter's gas analyzer. A comparison of gas concentrations for the three PURs under both ignition modes was made.

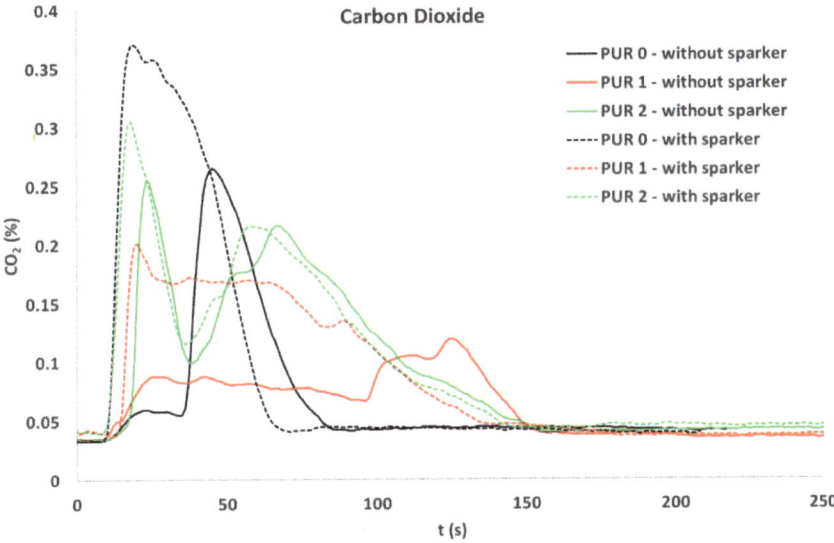

Figure 10. Carbon dioxide release, during exposure of the PURs to the 40 kW/m² of heat flux; self-ignited and ignited with a spark igniter.

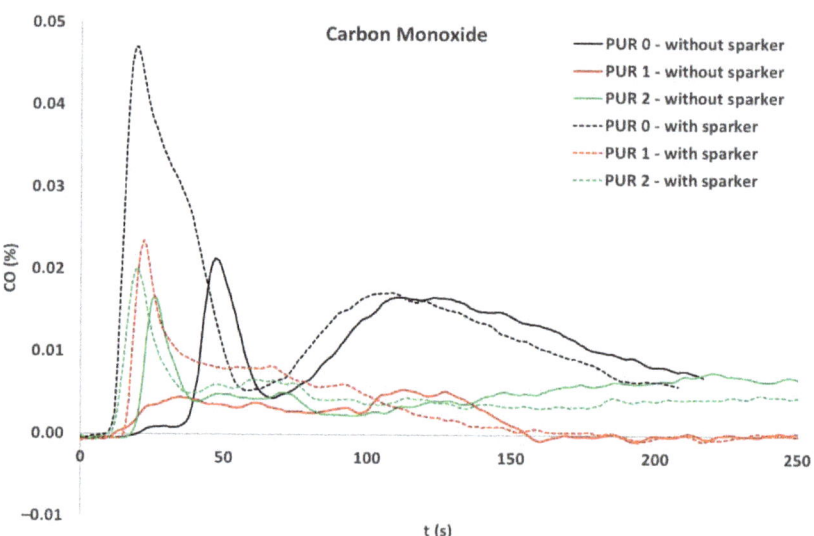

Figure 11. Carbon monoxide release, during exposure of the PURs to the 40 kW/m² of heat flux; self-ignited and ignited with a spark igniter.

During flaming both CO_2 and CO were released. After the flame was extinguished, the concentration of CO increased significantly for PUR 0.

In addition, NO was released during flaming in all tested specimens, while NO_2 was significantly noticed in PUR 0 and PUR 1 self-ignited specimens, which is clearly visible in Figures 12 and 13.

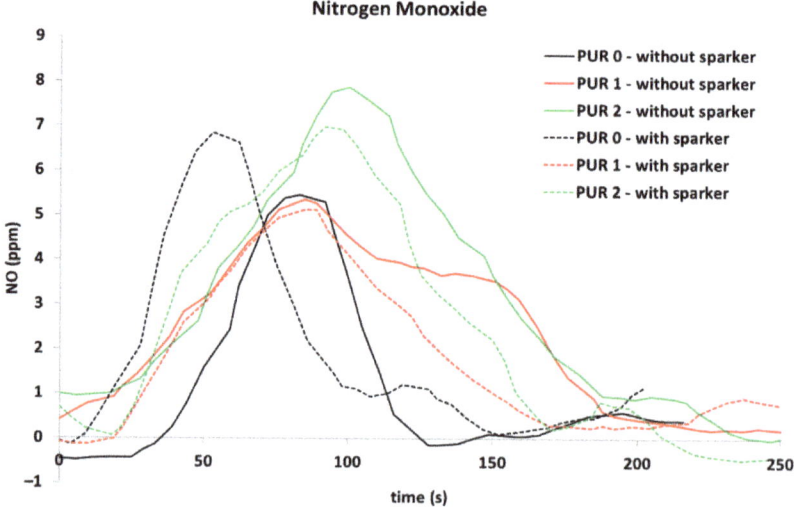

Figure 12. Nitrogen monoxide release, during exposure of the PURs to the 40 kW/m² of heat flux; self-ignited and ignited with a spark igniter.

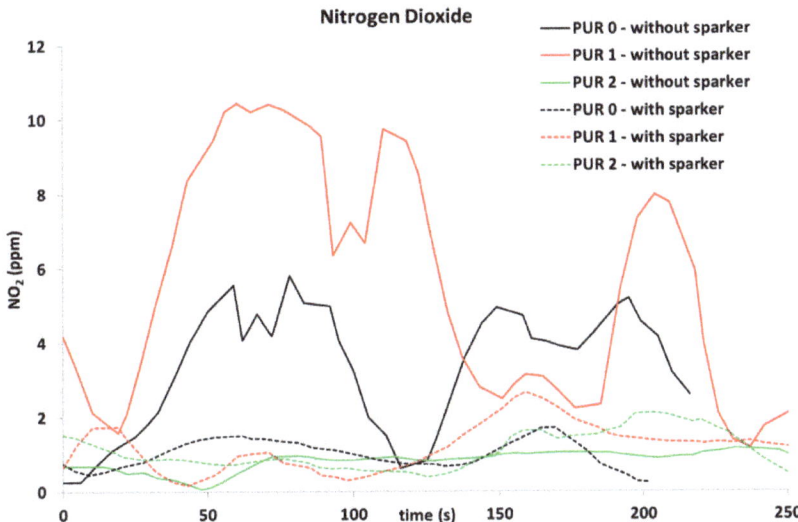

Figure 13. Nitrogen dioxide release, during exposure of the PURs to the 40 kW/m^2 of heat flux; self-ignited and ignited with a spark igniter.

We also detected the release of ammonia, which can be seen in Figure 14.

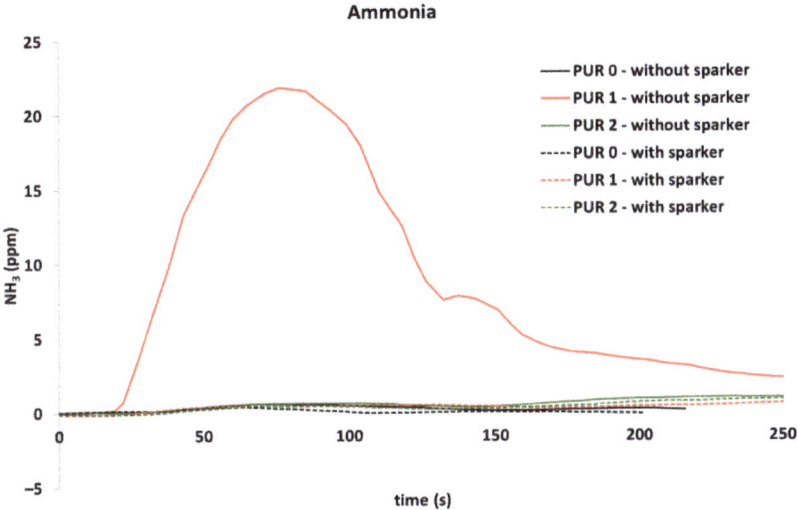

Figure 14. Ammonia release, during exposure of the PURs to the 40 kW/m^2 of heat flux; self-ignited and ignited with a spark igniter.

During exposure of the PUR 1 specimen to heat flux without sparks, some gases were detected, namely NH_3, C_2H_4, and C_2H_6, which were not seen in other specimens. Only in PUR 0 specimen, ignited with a spark, was C_2H_4 released also. When comparing self-ignited PUR 1 with other specimens it was noticed that the heat release of that specimen was significantly lower than with other specimens. It is possible that the flow of the chemical reaction was different. For finding the cause, some further investigations would be needed.

3.10. Surface Morphology

Monitoring the integration of the fire-retardant powder and the porosity of the materials was key in the SEM analysis [29,30]. In micrographs (Figures 15–17) of material surfaces with included fire-retardant powders are shown. The images were taken before (Figure 15) and after the addition of the fire-retardant powder (Figures 16 and 17) at 30×, 50×, and 100× magnification.

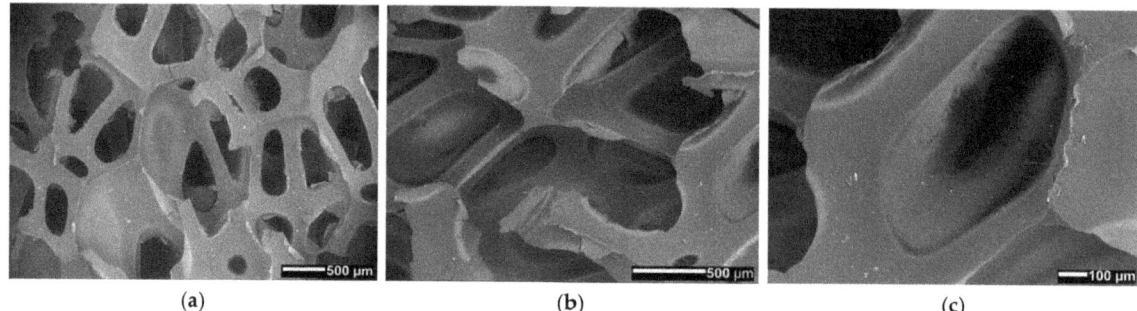

Figure 15. SEM micrographs of the surface of expanded solid polymer PUR 0 sample; at magnification (**a**) 30×, (**b**) 50×, and (**c**) 100×.

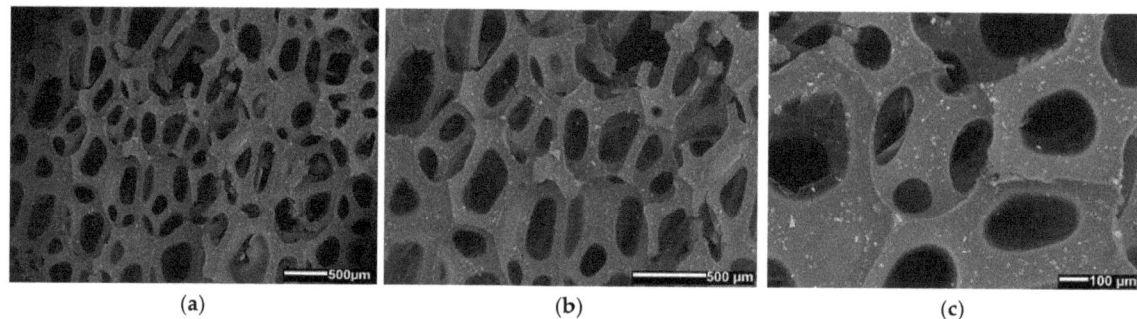

Figure 16. SEM micrographs of the surface of expanded solid polymer PUR 1 sample; at magnification (**a**) 30×, (**b**) 50×, and (**c**) 100×.

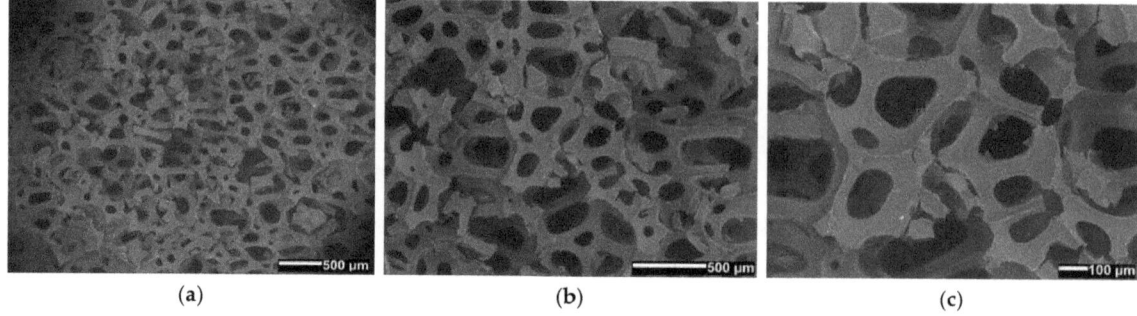

Figure 17. SEM micrographs of the surface of expanded solid polymer PUR 2 sample; at magnification (**a**) 30×, (**b**) 50×, and (**c**) 100×.

It can be seen from Figures 15–17 that PUR 0, without the addition of fire-retardant powder, forms the polymer network with the largest pores. In Figure 16 we can see in all three PUR 1 micrograms (a–c) that at the time of formation of the polymer network of the three-component PUR 1 composite, smaller pores were formed, which can be attributed to the addition of APP, and similarly, in Figure 17, where TATA was added as a fire-retardant powder, we can observe in PUR 2 micrograms (a–c) that even smaller pores were formed. The pores in all three prepared samples PUR 0–PUR 2 were random but evenly distributed. Figure 16 also shows many more solid particles on the surface than can be seen in Figures 15 and 17. We assume that the particles in Figure 16, which are pure two-component resin (binder + hardener) without additives, are possibly residues of an unreacted component or impurities left after sample preparation/cutting for SEM analysis. However, the significantly higher number of solid particles in Figure 16 may also be attributed to the fact that the APP may be more difficult to mix with the selected resin system, thus making the inhomogeneity of the PUR 1 sample much worse. Evaluation analysis of the agglomerate formation of micrographs was not performed because the powder is integrated also within the cavities of foamed materials, which made evaluation unreliable.

4. Discussion

As a thermal insulation material, PURs shall meet the demand of flame resistance, and at the same time also need to possess the necessary physical-mechanical properties. In this work, the influences of fire retardants of ammonium polyphosphates and 2,4,6-triamino-1,3,5-triazine on the properties of 2-component polyurethane foam were investigated.

Even though the additions of flame-retardant specimens contribute to increasing the density of PUR 1 and PUR 2 specimens their compressive strengths and corresponding strains were lower as compared to the PUR 0 reference. In the case of PUR 1, a 33% decrease in compressive strength and an 11% decrease in bending strength were recorded. For PUR 2 about a 7% decrease in compressive and a 14% in bending strengths were observed. It can be concluded that retardants are not chemically bonded into PU binders. However, it is worth pointing out that mechanical properties are not considerably affected by the additions of retardants. Thus, PUR 1 and PUR 2 still exhibited relatively high mechanical properties as compared to typical cellular insulation to be used for thermal insulation of the buildings.

The shore durometer hardness value itself does not provide direct information on, e.g., strength or resistance to scratches, abrasion, or wear. This hardness is a measure of a material's resistance to localize the plastic deformation, and it can be defined as a measure of a material's resistance towards an external force applied to the material. From the results, we can see that the addition of fire-resistant powders influenced resin network forming and thus decrease the durometer hardness values, even by 50%.

It was determined that presented retardants considerably decrease the flammability of the systems, while thermal conductivities are not affected.

The fire behavior of different PUR specimens depends on fire conditions. In conditions of starting a fire as in UL 94 standard, additives in PUR successfully reduce fire spread. In conditions toward the fully developed fire as simulated with the cone calorimeter test, specimens react differently. Specimens with additives compared to PUR 0 specimens burn longer, the heat release rate is lower, but total heat release is higher. The smoke production rate for specimens with additives is lower compared to PUR 0, also total smoke production is lower, especially for PUR 2 specimens.

When comparing loss on ignition, PUR 0 specimens had the highest LOI value, around 95%. PUR 1 specimens had the lowest LOI value, around 65%.

When observing the release of gases during heat exposure, it was noticed that CO significantly increased after the end of flaming at PUR 0 specimens, while in other specimens CO was not observed after the end of flaming. NO_2 and C_2H_4 were observed in PUR 0 self-ignited specimen and PUR 1 specimen, ignited by sparks. NH_3 and C_2H_6 were observed only at heat exposure of PUR 1 specimen, when self-ignited. The different gas releases can indicate different chemical reactions and can relate to the low heat release rate of that specimen during heat exposure. For a better understanding of the chemical processes, it would be necessary to do new, in-depth research related to the type of test, the variations of the test parameters, the amount of fire-retardant additives, etc.

In addition, it was found that the foamed PUR materials without and with integrated different fire-retardant powders have different shapes and sizes of porous or polymeric net structures.

The above research was carried out based on three different composites, and we should mention that we did preliminary research with different amounts of fire-retardant material additives, i.e., 10% and 30%, and 50%. At 10% addition, there were no significant differences in the UL94 HB test, while at 50% addition, we failed to homogeneously mix such a large amount of dust into the B component, so we selected samples with 30% addition and re-prepared the entire batch of 30% samples for conditioning and further analysis. The present work is the basis for a further, more detailed study of polyurethane systems, in which additional materials affecting the response to fire will be investigated, different concentrations of additives or different compositions of the systems will be investigated, as well as different methods of installation procedures and the effect on their mechanical properties.

Author Contributions: Conceptualization, B.M.; methodology, B.M. and J.B.; validation, B.M.; formal analysis, J.B., N.K. and B.M.; investigation, J.B., N.K. and B.M.; resources, B.M., N.K. and J.B.; writing—original draft preparation, B.M., N.K.; writing—review and editing, B.M. and J.B.; visualization, B.M., N.K. and J.B.; supervision, B.M.; project administration, B.M. All authors have read and agreed to the published version of the manuscript.

Funding: This research was funded by Slovenian Research Agency (ARRS), Slovenia; ARRS program group no. P2-0273.

Institutional Review Board Statement: Not applicable.

Data Availability Statement: Not applicable.

Conflicts of Interest: The authors declare no conflict of interest.

References

1. Gatto, M.P.A.; Lentini, V.; Castelli, F.; Montrasio, L.; Grassi, D. The Use of Polyurethane Injection as a Geotechnical Seismic Isolation Method in Large-Scale Applications: A Numerical Study. *Geosciences* **2021**, *11*, 201. [CrossRef]
2. Amaral, C.; Silva, T.; Mohseni, F.; Amaral, J.S.; Amaral, V.S.; Marquesa, P.A.A.P.; Barros-Timmons, A.; Vicente, R. Experimental and numerical analysis of the thermal performance of polyurethane foams panels incorporating phase change material. *Energy* **2021**, *216*, 119213. [CrossRef]
3. Gunther, M.; Lorenzetti, A.; Schartel, B. Fire phenomena of rigid polyurethane foams. *Polymers* **2018**, *10*, 1166. [CrossRef] [PubMed]
4. Chen, K.P.; Cao, F.; Liang, S.E.; Wang, J.H.; Tian, C.R. Preparation of poly(ethylene oxide) brush-grafted multiwall carbon nanotubes and their effect on morphology and mechanical properties of rigid polyurethane foam. *Polym. Int.* **2018**, *67*, 1545–1554. [CrossRef]
5. Wang, X.; Lijun, Q.; Linjie, L. Flame Retardant Behavior of Ternary Synergistic Systems in Rigid Polyurethane Foams. *Polymers* **2019**, *11*, 207. [CrossRef]
6. Bunderšek, A.; Japelj, B.; Mušič, B.; Rajnar, N.; Gyergyek, S.; Kostanjšek, R.; Kranjc, P. Influence of Al(OH)$_3$ nanoparticles on the mechanical and fire resistance properties of poly(methyl methacrylate) nanocomposites. *Polym. Compos.* **2016**, *37*, 1659–1666. [CrossRef]
7. Wu, W.; Zhao, W.; Gong, X.; Sun, Q.; Cao, X.; Su, Y.; Yu, B.; Li, R.K.Y.; Vellaisamy, R.A.L. Surface decoration of Halloysite nanotubes with POSS for fire-safe thermoplastic polyurethane nanocomposites. *J. Mater. Sci. Technol.* **2022**, *101*, 107–117. [CrossRef]
8. Usta, N. Investigation of fire behavior of rigid polyurethane foams containing fly ash and intumescent flame retardant by using a cone calorimeter. *J. Appl. Polym. Sci.* **2012**, *124*, 3372–3382. [CrossRef]
9. Heng, Z.; Shi-ai, X. Preparation and fire behavior of rigid polyurethane foams synthesized from modified urea–melamine–formaldehyde resins. *RSC Adv.* **2018**, *32*, 17879–17887. [CrossRef]
10. Hejna, A.; Haponiuk, J.; Piszczyk, Ł.; Klein, M.; Formela, K. Performance Properties of Rigid Polyurethane-Polyisocyanurate/Brewers' Spent Grain Foamed Composites as Function of Isocyanate Index. *Polymers* **2017**, *17*, 427–437. [CrossRef]
11. Amran, U.A.; Salleh, K.M.; Zakaria, S.; Roslan, R.; Chia, C.H.; Jaafar, S.N.S.; Sajab, M.S.; Mostapha, M. Production of Rigid Polyurethane Foams Using Polyol from Liquefied Oil Palm Biomass: Variation of Isocyanate Indexes. *Polymers* **2021**, *13*, 3072. [CrossRef] [PubMed]

12. Fink, J.K. Poly(Urethane)s. In *Reactive Polymers: Fundamentals and Applications*; Elsevier: Amsterdam, The Netherlands, 2018; pp. 71–138.
13. Kandola, B.K.; Horrocks, S.; Horrocks, R.A. Evidence of interaction in flame-retardant fibre intumescent combinations by thermal analytical techniques. *Thermochim. Acta* **1997**, *294*, 113–125. [CrossRef]
14. Ullah, S.; Ahmad, F. Effects of zirconium silicate reinforcement on expandable graphite based intumescent fire retardant coating. *Polym. Degrad. Stab.* **2014**, *103*, 49–62. [CrossRef]
15. Devallencourt, C.; Saiter, J.M.; Fafet, A.; Ubrich, E. Thermogravimetry/Fourier transform infrared coupling investigations to study the thermal stability of melamine formaldehyde resin. *Thermochim. Acta* **1995**, *259*, 143–151. [CrossRef]
16. Kaur, R.; Kumar, M. Addition of anti-flaming agents in castor oil based rigid polyurethane foams: Studies on mechanical and flammable behaviour. *Mater. Res. Express* **2020**, *7*, 015033. [CrossRef]
17. Yang, R.; Hu, W.; Xu, L.; Song, Y.; Li, J. Synthesis, mechanical properties and fire behaviours of rigid polyurethane foam with a reactive flame retardant containing phosphazene and phosphate. *Polym. Degrad. Stab.* **2015**, *122*, 102–109. [CrossRef]
18. Kandelbauer, A.; Tondi, G.; Goodman, S.H. Unsaturated Polyesters and Vinyl Esters. In *Handbook of Thermoset Plastics*, 3rd ed.; Dodiuk, H., Goodman, S.H., Eds.; William Andrew: San Diego, CA, USA, 2013; Volume 6, pp. 111–172.
19. Suoware, O.; Edelugo, S.O.; Ugwu, B.N.; Amula, E.; Digitemie, I.E. Development of flame retarded composite fibreboard for building applications using oil palm residue. *Mater. Construcción* **2019**, *69*, e197. [CrossRef]
20. Bachtiar, E.V.; Kurkowiak, K.; Yan, L.; Kasal, B.; Kolb, T. Thermal Stability, Fire Performance, and Mechanical Properties of Natural Fibre Fabric-Reinforced Polymer Composites with Different Fire Retardants. *Polymers* **2019**, *11*, 699. [CrossRef]
21. Arjmandi, R.; Ismail, A.; Hassan, A.; Abu Bakar, A. Effects of Ammonium Polyphosphate Content on Mechanical, Thermal and Flammability Properties of Kenaf/Polypropylene and Rice Husk/Polypropylene Composites. *Constr. Build. Mater.* **2017**, *152*, 484–493. [CrossRef]
22. Luo, F.; Wu, K.; Lu, M.; Nien, S.; Guan, X. Thermal degradation and flame retardancy of microencapsulated ammonium polyphosphate in rigid polyurethane foam. *J. Therm. Anal. Calorim.* **2015**, *120*, 1327–1335. [CrossRef]
23. Zhao, X.L.; Chen, C.K.; Chen, X.L. Effects of Carbon Fibers on the Flammability and Smoke Emission Characteristics of Halogen-Free Thermoplastic Polyurethane/Ammonium Polyphosphate. *J. Mater. Sci.* **2016**, *51*, 3762–3771. [CrossRef]
24. Carosio, F.; Alongi, J.; Malucelli, G. Layer by Layer Ammonium Polyphosphate-Based Coatings for Flame Retardancy of Polyester-Cotton Blends. *Carbohydr. Polym.* **2012**, *88*, 1460–1469. [CrossRef]
25. Vitkauskiene, I.; Makuška, R.; Stirna, U.; Cabulis, U. Thermal Properties of Polyurethane-Polyisocyanurate Foams Based on Poly(ethylene terephthalate) Waste. *Mater. Sci.* **2011**, *17*, 249–253. [CrossRef]
26. Price, D.; Liu, Y.; Hull, T.R.; Milnes, G.J.; Kandola, B.K.; Horrocks, A.R. Burning behaviour of fabric/polyurethane foam combinations in the cone calorimeter. *Polym. Int.* **2000**, *49*, 1153–1157. [CrossRef]
27. Aydogan, B.; Usta, N. Cone calorimeter evaluation on fire resistance of rigid polyurethane foams filled with nanoclay/intumescent flame retardant materials. *Res. Eng. Struct. Mat.* **2018**, *4*, 71–77. [CrossRef]
28. Tang, G.; Zhou, L.; Zhang, P.; Han, Z.; Chen, D.; Liu, X.; Zhou, Z. Effect of aluminum diethylphosphinate on flame retardant and thermal properties of rigid polyurethane foam composites. *J. Therm. Anal. Calorim.* **2020**, *140*, 625–636. [CrossRef]
29. Thirumal, M.; Khastgir, D.; Singha, N.K.; Manjunath, B.S.; Naik, Y.P. Effect of Expandable Graphite on the Properties of Intumescent Flame-Retardant Polyurethane Foam. *J. Appl. Polym. Sci.* **2008**, *110*, 2586–2594. [CrossRef]
30. Buzzi, O.; Fityus, S.; Sasaki, Y.; Sloan, S. Structure and properties of expanding polyurethane foam in the context of foundation remediation in expansive soil. *Mech. Mater.* **2008**, *40*, 1012–1021. [CrossRef]

www.ingramcontent.com/pod-product-compliance
Lightning Source LLC
LaVergne TN
LVHW070644100526
838202LV00013B/878

Disclaimer/Publisher's Note: The statements, opinions and data contained in all publications are solely those of the individual authors(s) and contributor(s) and not of MDPI and/or the editor(s). MDPI and/or the editor(s) disclaim responsibility for any injury to people or property resulting from any ideas, methods, instructions or products referred to in the content.

MDPI
St. Alban-Anlage 66
4052 Basel
Switzerland
www.mdpi.com

Polymers Editorial Office
E-mail: polymers@mdpi.com
www.mdpi.com/journal/polymers